STOREY'S GUIDE TO
RAISING
PIGS

STOREY'S GUIDE TO
RAISING
PIGS

Kelly Klober

STOREY
BOOKS
Schoolhouse Road
Pownal, Vermont 05261

*The mission of Storey Communications is to serve our customers
by publishing practical information that encourages
personal independence in harmony with the environment.*

Edited by Elizabeth McHale and Diana Delmar
Cover design by Susan Bernier
Cover photograph by Grant Heilman Photography, Larry Lefever
Back cover photograph by PhotoDisc
Production by Susan Bernier and Erin Lincourt
Line drawings by Becky Turner
Indexed by Word•a•bil•i•ty

Printed in the United States by Versa Press
10 9 8 7 6 5 4 3 2

Library of Congress Cataloging-in-Publication Data

Klober, Kelly, 1949–
    [Guide to raising pigs]
    Storey's guide to raising pigs / Kelly Klober.
       p. cm.
    Originally published: Guide to raising pigs. Pownal, Vt. : Storey Communications, c1997.
    Includes index.
    ISBN 1-58017-326-8 (alk. paper)
    1. Swine. I. Title.

SF395 .K54 2000
636.4—dc21
                                         00-030804

CONTENTS

DEDICATION

To my wife, Phyllis Klober,
and my grandparents, Kelly and Elizabeth Brewer,
who helped to make all my farming dreams come true.

ONE

AN INTRODUCTION TO RAISING HOGS

OF ALL OF THE MAJOR LIVESTOCK SPECIES, none is more misunderstood or less appreciated than the hog.

The rooting, squealing, twisted-tailed mortgage lifter of the Midwest is actually known and valued far beyond the borders of the Corn Belt and the rim of the breakfast plate. The ancient Egyptians sowed seeds into ground broken and trampled by the sharply pointed, cloven hooves of hogs. There is evidence of their domestication reaching back as far as six thousand years, making them a contender for the title of the oldest of the domestic species.

Hogs arrived in the New World with the conquistadors and moved ever westward with the expanding frontier. Along the American frontier, cured pork products — such as hams and bacons — even served as an early form of currency.

In the classic folk song "Sweet Betsy from Pike," a spotted hog figured quite prominently in the livestock inventory that Betsy and her husband, Ike, moved west with them. The Pike in that song title was Pike County, Missouri, which to this day remains one of the best-known pork-producing counties in the nation. The first Duroc gilt I bought, nearly 30 years ago, hailed from Pike County, too.

The strength of those early ridge runners and razorbacks continues. No major livestock species is more hardy, adaptable, or productive than the hog. Here at Willow Valley, Missouri, I have had sows produce three litters in a single 12-month period. An average of 2.25 litters per sow per year, with 7 or 8 pigs per litter, is doable on even the smallest of farms. There are sows on moderate-size commercial farms in both Europe and the United States that are producing nearly 26 pigs each per year.

Small producers with simple facilities following a seasonal plan of production can realize two litters per sow per year without taxing their time or resources. And it is reasonable to expect those litters to have a weaning average of eight or even nine good pigs.

Modern Pork Production

The pigs of today are efficient converters of feedstuffs to lean, healthful meat. On average, the pork produced today is as much as 30 percent leaner than the pork of the 1950s, when much of the human nutritional data that is still in use was gathered. Through selective breeding, better management, and improved feeding practices, modern pork has become a source of high-quality

Most hogs as we know them today are thought to have descended from the Eurasian wild boar. From Asia, they spread to Europe and Africa.

protein and such valuable nutrients as iron and zinc, with a fat content comparable to chicken in the same-size serving portions.

Lean or Lard?

The modern hog is a meat animal of truly extraordinary abilities. In fact, a distinction is no longer made between bacon or meat-type hogs and lard hogs. There simply are no more of the latter category, and even those breeds

Serving Sizes and Nutritional Profiles of Lean Meats

3 ounces of cooked, roasted, trimmed	Calories	Total Fat (g)	Saturated Fat (g)	Cholesterol (mg)
Lean Chicken				
Skinless chicken breast	140	3.1	0.9	73
Skinless chicken leg	162	7.1	2.0	80
Skinless chicken thigh	178	9.3	2.6	81
Lean Cuts of Pork (*roasted, **broiled)				
Pork tenderloin*	139	4.1	1.4	67
Pork boneless sirloin chop**	164	5.7	1.9	78
Pork boneless loin roast*	165	6.1	2.2	66
Pork boneless top loin chop**	173	6.6	2.3	68
Pork loin chop**	171	6.9	2.5	70
Pork boneless sirloin roast*	168	7.0	2.5	73
Pork rib chop**	186	8.3	2.9	69
Pork boneless rib roast*	182	8.6	3.0	70
Lean Cuts of Beef (*roasted, **broiled, ***braised)				
Beef eye of round*	141	4.0	1.5	59
Beef top round***	169	4.3	1.5	76
Beef tip round*	149	5.6	1.8	669
Beef top sirloin**	162	5.8	2.2	76
Beef top loin**	168	7.1	2.7	65
Beef tenderloin**	175	8.1	3.0	71
Fish (*dry heat, **moist heat)				
Cod*	89	0.7	0.1	40
Flounder*	99	1.3	0.3	58
Halibut*	119	2.5	0.4	35
Orange roughy*	75	0.8	0.0	22
Salmon*	175	11.0	2.1	54
Shrimp**	84	0.9	0.2	166

National Pork Producers Council

and lines strongest in the genetic traits for litter size and mothering ability produce trim carcasses of exceptional merit. It has been more than 20 years since I have seen a butcher hog that could be legitimately termed a No. 4 butcher (lardy).

There are now boars approaching that fabled 2:1 feed efficiency ratio at which pork rivals poultry in production costs. Carcasses are getting longer, with a great many now exceeding 33 inches in length; loineyes are increasing to well past 6 square inches; and plate-size pork chops are being achieved through bigger loins, not smaller plates. Many market hogs now exceed 70 percent in the amount of lean they yield when dressed. A few years back we had a gilt slaughtered that was so lean we had to buy fat to add to her trim to make sausage with the desired cooking qualities.

> **Space Requirements**
>
> Space for a pig or two being fed out for the table can often be measured in mere square feet. One-quarter acre of pasture can carry up to four nursing sows and their litters. Few are the smallholdings that can't support a small swine venture of one sort or another.

Pork Today

Today's pork comes in the same popular forms it did years ago. Bacon, ham, sausage, and Canadian bacon still have a solid claim on that first and most important meal of the day, breakfast. The ham sandwich remains one of the leading sellers in the restaurant trade, and new products such as pork burger have achieved great success. Baked ham and pork chops are still among the all-time favorite comfort foods.

But there are some differences between the pork of the 1950s and modern pork. Ounce for ounce, pork today is about 30 percent leaner. It has a fat content comparable with chicken, while providing high-quality protein, iron, and zinc. Check out the chart (on page 3), from the National Pork Producers Council.

The Pork Producer

Hogs have a strong midwestern identity, but they are actually produced from coast to coast by small and midsize producers operating in climates and regions as varied as Maine and Hawaii. There is indeed a role for pork producers of all sizes including those with small-size sow herds. Our own herds have generally run from just 5 to 10 head. Our veterinarian, who is based in Pike County, judges the average sow herd in his practice to number just 35 head, and that's in one of the leading pork-producing counties in Missouri.

There is a great misconception that you must have a large acreage and hogs by the hundreds for production to be a viable enterprise. Nothing could be farther from the truth. We once sold two feeder pigs to a couple who asked if their 20-acre pasture at home would be large enough to accommodate them. On that 20 acres there was certainly room for those two shoats — along with several thousand more, if housed correctly. On our own 2.86 acres we have raised as many as 10 sows and their offspring as well as maintaining several other small livestock and poultry ventures.

Hog Myths

Since I've started myth busting, let me address several other mistaken beliefs about hogs and hog raising:

Hogs are inherently dirty animals. This belief just isn't true. Hogs do lack the ability to sweat; hence they sometimes run into problems in very hot and humid weather, because they can release heat only through panting and the evaporative cooling process. They will therefore wallow in muddy water to cool off. Otherwise, though, they have no real affinity for mud. In small lots and around feeding and watering equipment, their sharply pointed hooves do tear at the soil surface and can cause muddy areas to develop.

Their rooting activity in the quest for mast (roots, fallen fruits and nuts, grubs, and other natural feedstuffs found at or just below the soil's surface) also muddies their image. This is an instinctive feeding process dating back to their feral ancestors, which were denizens of the forest floors of Europe. A hog on pasture or in a wooded lot can still do much to balance its own diet, although it is no longer necessary to provide pasture and browse for hogs.

Hog Handling

When handling hogs in close quarters, just as with any other large animals, plan ahead and leave yourself at least two ways out. Move about hogs with a quiet and assured manner. Never back them into a corner — they may have no choice but to strike out by lunging or biting.

Hogs are inherently greedy or gluttonous creatures. The reason they are mistakenly assumed to be gluttonous is that young and growing hogs are fed to full appetite (fed to consume roughly 3 percent of their bodyweight daily in feedstuffs) to foster rapid growth to market weight. It is also part of their herd instinct for all of the hogs in a group to rise up and want to eat at the same time. Further, most sows are fed limited quantities throughout much of their gestation period; if a sow amasses less fat, she carries the litter more easily, and the delivery is less stressful. This limited feeding explains why sows are eager to eat when feed is presented to them.

Hogs are very efficient users of feedstuffs, often averaging 1 pound of gain on just 3 to 3.5 pounds of feedstuffs, even in the simplest of facilities.

Also, as noted earlier, they are accustomed to rooting about for at least a portion of their diet, but they are not eaters of swill and garbage as so often depicted. In earlier days they would root through the streets and alleys of rural villages, but their quest was for table scraps, mill wastes, and spilled grains.

The base ingredient in most modern swine rations is one of the coarse feed grains such as corn, grain sorghum, or barley, and the most commonly used protein supplement is soybean oil meal. The corn is not the sweet corn we humans enjoy but yellow field corn, which has a quite low crude protein content. It has never figured greatly in the human diet. Those who think that hogs are crude eaters should bear in mind that they have a digestive system quite similar to our own, are often used in human-related medical research, and, other than humans, are the only creatures that will consistently and will-ingly consume alcohol for study purposes.

You have to raise and sell swine in huge numbers to make money. There are few livestock species that present as many marketing opportunities as hogs: butcher stock, feeder pigs, crossbred and purebred breeding stock, roasting pigs, direct sale of meat animals and pork, and more — many more. Pork sausage from our simply reared hogs easily brings $2 a pound, our young boars sell in the $250 to $300-per-head range, and our surplus pigs retain good value as feeder stock. The key is whether you are practicing good marketing or merely dumping your production on whatever outlet is available at the moment.

Hogs are more vicious or treacherous to work with than other livestock species. Granted, one of our 10-pound rabbits is easier to handle than one of our 450-pound sows on a one-to-one basis, but over the years the rabbits have bestowed far more scrapes and scratches. With hogs you simply have to build housing and handling facilities that are sturdy enough for such large animals. Housing for hogs should be reinforced at the ground and at hog height, where the animals are, not at the head height of a human. Further, anyone working with hogs should respect them for their strength and size, and for their often surprising quickness and agility.

The tales of mean hogs date back to the times of free range, when you were as likely to encounter feral hogs as domestic ones. If you were attacked, the standard advice was to tuck in your head, cover it with your arms, and draw up your knees to your body. We worry about this less today, because for many generations now hogs have been selectively bred for docility and a quiet temperament.

Modern Swine Breeds

The three most popular breeds of swine in the United States now are the Duroc, Hampshire, and Yorkshire. They and their crosses are the backbone of the pork industry. Their development and preservation was the loving work of generations of small and midsize family farmers.

A distinction of sorts is now made between colored and white breeds of swine. Although all of today's swine breeds are selectively bred for leanness, efficiency, and meatier carcasses, the colored breeds are still considered to have better "economic" traits. As a group they are noted for their vigor, faster yet leaner growth, and meatier carcasses.

The white breeds, on the other hand, are strong in the traits needed for successful pig raising. They milk better than colored breeds and as a group, tend to farrow larger litters. They also have the docile nature you need when you are raising and weaning large litters. On commercial farms the goal is to blend these genetics — and thus these breed-specific traits — to produce goodly numbers of hogs that possess the best qualities of the different breeds.

Colored Breeds

The Duroc is a truly American breed. It was originally known as the Duroc-Jersey, which was a cross between two types of hogs: the Jersey Red of New Jersey and the Duroc of New York. The Duroc also is red in color, ranging from a dark, brick red to several lighter shades. It has smallish, drooping ears.

The Duroc is often used as a terminal cross male in a crossbreeding program for its muscling and meat qualities. *Terminal* means a cross to maximize growth and muscling, since all pigs produced are ultimately butcher stock. Duroc pork is considered among the best for eating qualities. Duroc hogs also are among the hardiest of the swine breeds and are widely used in outdoor operations. Duroc females are reasonably good mothers, and the gilts from crosses with Hampshires and Yorkshires work really well outside and in simple housing.

History of the Duroc

A lively history of the Duroc can be found in *Modern Breeds of Livestock, 1969,* Third Edition, by Hilton M. Briggs. It seems that a man from Saratoga County, New York, named Isaac Frink obtained his first hogs in the 1820s from Harry Kelsey. It was Mr. Kelsey who owned a famous thoroughbred stallion named Duroc. While Mr. Frink was visiting Mr. Kelsey, he spied red pigs that took his fancy, so he bought some and took them home. The pigs had no breed name so he called them Durocs, in honor of the magnificent stallion.

The Jersey Red was well established in New Jersey by the mid-1850s. These hogs were extremely large, rugged, and prolific. But they lacked in quality because they had coarse haircoats and bone, and were long and rangy.

It was the crossing of these two breeds that led to today's Duroc. Interestingly, however, the breed was built not on the East Coast where it originated, but in the Midwest.

The Hampshire is — as are all of the breeds with *shire* in their name — of English origin. It is a black hog with a distinctive white belt encircling it in the shoulder region. This color pattern dates to a fad for breeding all manner of domestic animals for this striking color effect. Animals without a white belt encircling the whole body, or white splashes in the upper part of the body, will not be issued a pedigree document. These are often termed "off belts" and are sold at somewhat discounted prices. They are not impaired from producing quality meat animals in any way.

With erect ears and a trim appearance, the Hampshire is very alert and vibrant in appearance. Many producers favor hogs with erect ears, believing them to be easier to handle and drive. I believe that you can tell more about a hog by observing its temperament than its ears, but I will admit that some hogs with erect ears seem easier to handle and move "straight away."

Hampshire boars are also often used as terminal sires, and the Hampshire x Yorkshire gilt is the stuff of legend in the swine industry for its hardiness and productivity. Hampshires impart rapid growth and exceptional leanness to their offspring.

Hampshire purebreds do not, on average, produce especially large litters, but the Hampshire sow that we used to produce Hamp x Duroc boars maintained a litter weaning average of 11 for eight litters. Many in the swine industry believe that the milk produced by a nursing Hampshire sow is nutritionally denser than that of many other breeds.

History of the Hampshire

Hampshires are thought to be one of the oldest original breeds in the United States. According to the Hampshire Swine Registry, they probably originated from the "Old English Breed," a black hog with a white belt that was popular in Scotland. Hampshires came to America in the early 1800s from Hampshire county in England and were developed in Kentucky, where they were known for their hardiness, vigor, and foraging characteristics. They were also called McKay hogs because a man of that name was thought to have brought them from England to America. It was in the early 1900s that the Hampshire breed "swept across the corn belt like prairie fire," the swine registry says. Hampshires were also used to produce the famous Smithfield ham.

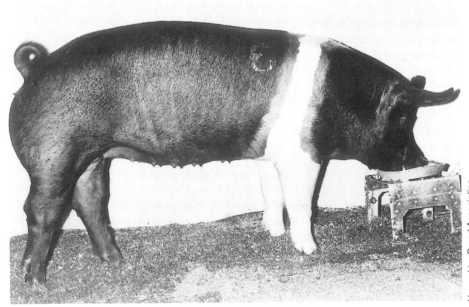

This Hampshire open gilt shows a lot of frame and muscling.

Hampshire Breed Association

The Spotted Breed

The Spotted breed offers a different choice for those who need both the hardiness of a black-colored breed and good litter size in their black/white/red, three-breed rotational crossbreeding plans. An example of such a cross would be a Spot/Duroc/York.

The Spotted, once known as the Spotted Poland China, is one of the colored breeds that have black and white spots. It is sometimes primarily black, and sometimes primarily white. This breed, which also has drooping ears, is known for its good mothering ability. Its meat quality is considered moderate: The carcass is long and trim but perhaps not as heavily muscled as the Hampshire or Duroc.

The Spotted is another of the old-line breeds down somewhat in its numbers from times past. Still, it is a breed with a lot of uses. Of the colored breeds, it produces some of the largest litters. The hogs are also of good length. For a time I produced Duroc x Spotted F_1 boars that were very well accepted by small producers seeking to add hardiness and litter size to their "outside" sow herds.

Spotted Boar Breed Association

This is a very rugged, powerful Spotted boar. He has massive bone — evidenced in his feet and legs — excellent rear leg placement, and a great topline. He's the classic "he-boar."

Spotteds share a common ancestry with a now minor British breed known as the *Gloucester Old Spot*. In the United States, the two are registered by the same breed association, into the same herd book. In England, the Old Spot enjoys support from the royal family. It is known as the orchard hog, and some claim that it was bred for over 100 generations totally outside and free of confinement.

The Berkshire is an old-line breed now enjoying a resurgence in popularity. In the 1930s it and the similarly colored Black Poland China led the nation in number of recorded pedigrees. It is a largely black breed with white points and splashes found predominantly in the face and lower half of the body. It has rather short, erect ears.

At one time the breed had a very short and upturned snout, but selective breeding in recent years has removed this trait, which many thought was conducive to feed waste and respiratory problems. The Berk is a well-muscled breed noted for both leanness and high-yielding carcasses. Pork from pure- and high-percentage bred Berkshires is noted for exceptional table qualities, including quite large loineye areas and finishing qualities that give the pork something akin to marbling in high-grade beef. This pork is especially valued in Japan, where it sells for a premium.

Berkshire Breed Association

This Berkshire boar is truly growthy and modern in his type. Note his trim yet masculine-appearing head.

Again, the Berkshire is often thought of more as a terminal sire. However, the females rank fairly high as mothers, and Berkshires are hardy and durable.

The Black Poland is nearly identical to the Berkshire in its color pattern, but has drooping ears. This hog has perhaps a flatter and longer topline than the Berkshire does, along with exceptional meat qualities: The carcasses are lean, are well muscled, and yield a good dressing percentage as well as hams.

The Black Poland also is widely known as one of the most durable of all of the swine breeds. At one time the Black Poland breed could lay claim to more females in its herd book at eighth parity and beyond than any of the other major swine breeds. Black Poland females are often underrated as brood sows. They can add real punch to a swine farm where hogs are kept outside on pasture or in a drylot outside and a very hardy black breed is needed to work into a crossbreeding rotation.

The Black Poland and Berkshire are both gaining in popularity as sires of show or youth project pigs. They have slipped sharply in numbers in recent years but have a bright future as breeds that have much to contribute to the production of lean, well-muscled market hogs.

Iowa Pork Industry Center

The Black Poland is one of the most durable of all of the swine breeds.

White Breeds

What is called "white" color in hogs is not really the color white, but the lightest shade in the brown color range. It is a dominant color in swine; in crossbreeding it may totally block out other colors, or allow them to appear only as a faint dappling on the rumps of the crossbred offspring. None of the white swine breeds are albinos.

In the United States, the white breeds are referred to as the "mother breeds." They are known for larger litter size and good milking ability, and they are proven raisers of large litters of pigs. Their overall conformation is even somewhat comparable with that of a dairy cow. They can and often do produce fairly high-yielding carcasses. Two of the meatiest boars I have seen in recent months were Yorkshires.

The Yorkshire lays claim to the title of the mother breed in its advertising and, in most years, leads the nation in total registrations. Large litter size at both birth and weaning is a breed standard and evidence of its superior mothering and milking abilities. A great many recommend this breed for a youngster considering a gilt and litter 4-H or Future Farmers of America (FFA) project, as they will provide the youngster with a great many pigs with

Yorkshire Breed Association

The Yorkshire is known as the mother breed. Notice the good underline, depth of side, and overall feminine appearance.

which to work. They are also among the most widely available of all of the purebred swine breeds.

The Yorkshire is all white, with sharply erect ears and dark eyes. It shares its origins with the *English Large White* — both are recorded with the same breed association in this country. The Large White is perhaps the definitive confinement breed.

Yorkshires also figure prominently in a great many of the more popular crosses, including some with the other white breeds, to supply highly productive female replacements for commercial operations. The Duroc x Yorkshire is growing in popularity in the Midwest as a way of creating an even more durable carrier for white genetics. The smallholder with simpler facilities will be best served by what are termed "American" Yorkshire breeding lines, which are more selectively bred for life in outside lots and pens.

Yorkshire History

The Yorkshire originated in England, and the first of these animals to cross the sea is thought to have been brought to Ohio around 1830. The American Yorkshire Club says that the breed initially failed with farmers because the hogs were slow growing and had short, pugged noses. A man in Indiana is credited with showing farmers the advantages of Yorkshire breeding stock, which include larger litters, great mothering ability, and more length. Once farmers realized these benefits, the breed gained in popularity. The first Yorkshire registered in the United States was Clover Crest, a boar imported from Canada.

Today, Yorkshires can grow at a rate of over 3 pounds per day, with an excellent feed conversion. The goal of the Yorkshire breed, says the Yorkshire club, is to be the source of durable mother lines, contributing to longevity and carcass merit.

The Landrace is considered the longest of all of the more popular swine breeds and is also known for its distinctive long, drooping ears. Selective breeding has greatly reduced Landrace ear size of late, however. This white breed also produces very large litters of pigs.

A bit finer boned than other breeds, the Landrace is more often used for indoor, confinement breeding than other breeds. Among the most docile of

Mike and Jackie Wagoner

The Landrace is a bit finer boned than other breeds and is often used for indoor, confinement breeding.

swine breeds, it is frequently used in crossbreeding to add to both litter size and carcass length to the resulting offspring. Separate Landrace strains are bred throughout the nations of Europe and Scandinavia, where its temperament is valued in the close-confinement housing often common there.

If penned outside, some Yorkshire and Landrace sows may need a bit of extra feed to maintain body condition. They also seem to perform best when held in groups of similar breeding.

The Chester White is an American-developed white breed that originated in Chester County, Pennsylvania. It used to be called the Chester County White. It has a medium-size frame and drooping ears. It is probably an underutilized white breed, considering its hardiness, which makes it appropriate for producers working outdoors with simple facilities.

The sows are good mothers and breed back very quickly following weaning, a very important economic consideration. Not only does this reduce operating costs, but there is also strong statistical evidence that the more quickly a sow breeds back following weaning, the larger her following litter will be. We have owned Chester sows that farrowed three litters in one 12-month period, and we have owned many that maintained lifetime litter averages of 10 or more weaned.

Chester White pigs are a bit smaller at birth than some others, but are generally quite vigorous. Consider, too, that with large litters, the pigs are more likely to be smaller. Purebred Chesters are also a bit slower growing than some other purebreds, but still turn in respectable days of age to 230 pounds. One of their real strengths — verified by university testing — is how and

Chester White Breed Association

This Chester White is a real sweetheart of a gilt. She's got a great underline and lots of body capacity. Notice the evidence of frame in her legs and shoulders.

Traits of Popular Breeds

Breed	Traits
Duroc	Top eating quality; hardy enough to raise outside
Hampshire	Offspring grow rapidly, with exceptional leanness
Berkshire	Premium table-quality pork
Black Poland	Especially durable; good for crossbreeding in an outdoor swine operation
Spotted	Excellent mothering ability, large litters
Yorkshire	The mothering breed; large litters; good for indoor confinement
Landrace	Large litters; docile; usually raised indoors
Chester White	Hardy white breed; breeds back quickly

what they can contribute in crossbreeding. They carry the typical strengths of the white breeds — milking, mothering, and litter size — while allowing the growth and carcass traits of other breeds in the cross to shine through.

Minor Breeds

Another group of swine breeds found in the United States are termed the minor breeds and, sad to say, some of them have to be considered truly endangered breeds. They are a valuable group because they include some of the hardiest of all swine genetics and are often among the leanest of the purebreds.

The Tamworth is a light red breed of English origin with erect ears. This is probably the most vigorous of the swine breeds and one that has an unearned reputation for having a bad temper. For instance, one of the breed association presidents who lived in our county regularly worked in pens with Tamworth sows and very young pigs, even though he was encumbered by a debilitating disease.

A bit smaller framed than some other breeds, the Tamworth has remained hardy and maintains many of the characteristics of what used to be called the "range hog." An acquaintance of mine tells of once unloading 10 Tamworth

Donald Bixby/American Livestock Breeds Conservancy

The Tamworth is probably the most vigorous of the swine breeds. Its reputation for having a bad temper is unearned.

sows and 101 pigs into a large wooded pasture; he left them with self-feeders, but they were otherwise largely unattended for several weeks. When he returned to gather them up at weaning age, the sows had safely raised all 101 pigs. The Tamworth sow lies down by first dropping to her front knees and then shuffling down in a way that keeps her from crushing pigs through overlay. In fact, Tamworth sows are deemed to be the best mothers of all of the colored breeds of swine.

The Tamworth was once used as an example for what used to be called the "bacon breed" of swine. It is a very lean hog, and some of its lines are too thin. Unfortunately, like many of the other minor breeds, it has become a bit stuck in time. Tamworths need more producers to take them up and selectively breed them — to maintain the traditional strengths of the breed but also to improve growth and muscling, which translate into better carcass yield and more cuts that are prime hams and loin.

The Wessex is a now rarely seen breed with the conformation of the Landrace, including the drooping ears, and the distinctive black and white color pattern of the Hampshire. Many believe it to be hardier than the Landrace; it tends to produce somewhat smaller litters, however. It is a valuable breed lacking in breeders and numbers. There may be more of these animals in Canada, although the Wessex is valued by many breeders in the United States who raise hogs outside on pasture or in drylots.

Another minor breed with a stronger Canadian presence, but sometimes heard about in the United States, is the *Large Black*. A prolific breeder and good pig raiser, this breed is of English origin. It is all black in color and has erect ears. Although it isn't as long or trim as other breeds, the Large Black represents some enduring swine genetics that could be brought back into use. Like other minor breeds, the Large Black has been neglected by the industry, supported only by a few breeders, and subjected to higher levels of close breeding; it endures, nevertheless, due to its innate vigor.

The Hereford is an especially attractive breed, with a red and white pattern mimicking the distinctive color pattern of the Hereford breed of cattle, and drooping ears. Its frame is a bit smaller than that of some other swine breeds.

I have seen purebred Hereford hogs place well in recent market hog shows. At one, a purebred Hereford gilt topped the always competitive class of 240- to 245-pound market hogs — the prime marketing weight for these hogs. Friend Ron Macher, publisher of *Small Farm Today* magazine, often buys purebred Herefords to feed out and process into extra-lean pork sausage, which he markets directly as a gourmet product.

The Mulefoot

A very minor and truly endangered swine breed is the *Mulefoot*. Today it exists only in the black color phase, has drooping ears, and has a solid hoof like a mule — not the cloven hoof with two toe points seen in other swine breeds.

One of the last herds of the Mulefoot breed endured because it was isolated on an island in the Mississippi River near Louisiana, Missouri. The herd lived basically on its own, and its survival is a testament to the hardiness of the breed. The Mulefoot is a bit fine boned and lacks modern meat qualities such as well-muscled, high-yield, long carcasses. It has, however, been taken up by the American Livestock Breeds Conservancy and some affiliated breeders. This should help assure that the Mulefoot will provide truly alternative genetics for the swine industry.

It was said that the Mulefoot breed was resistant to the once-dreaded swine disease hog cholera because it had no cloven hoof, which prevented the disease organism from entering the hog's body. This is a good story, but has absolutely no basis in fact.

The Mulefoot is a very hardy breed, perhaps closer to the wild hog than any other current swine breed. The red and spotted color phases of this old, old breed are now considered extinct. One text from the last century refers to it as an old breed of swine even then. When crossed with other hogs, the mulefooted trait might appear only on the front or back two feet, or not at all.

This is the Mulefoot, a truly endangered species.

Mark Fields/American Livestock Breeds Conservancy

The Red Waddle or **Wattle** is a near wild hog with a red color and fleshy wattles dangling from its jowls. It enjoyed a brief burst of popularity a few years ago when interest in exotic animals surged. Although some claimed that these hogs produced a "different" pork with leaner qualities, this was more likely due to limited feeding and slow growth. Because these claims didn't always hold up, interest in them slipped away almost as quickly as it arose.

Other feral and near feral strains and varieties are novelties or historically interesting, rather than breeds that can be practically used on a working farm. A few producers here and in England are experimenting with range systems to raise and market Wild and Russian boars as gourmet game meat and to stock game preserves. Wild hog populations, however, often harbor rather virulent disease organisms; also, feral hogs generally produce just one litter of four pigs or fewer per year, and those pigs are small and slow growing.

In the last one hundred years, at least eight pure breeds of swine have become extinct in this country. The last disappeared in my lifetime: The **Ohio Improved Chester,** or O.I.C., was absorbed by the Chester White breed. Other breeds, such as the **Sapphire** and **Red Berkshire,** simply fell out of favor or failed to have the numbers adequate to keep up with changes in swine type. A swine text I own from very early in this century lists populations of a number of breeds that no longer exist and then, in a most tantalizing note, contains a heading for breeds even then existing in numbers too small to list.

Why Purebred Hogs?

There are not as many purebred hogs today as there were even 20 years ago. I recently came upon a farm magazine published the year after I graduated from high school — 1969. In it was page after page of advertising for purebred breeders that have slipped away just within the last decade. In that year, there were at least 15 purebred spring and fall auctions within 50 miles of our home.

Those purebred animals went sometimes to other purebred producers, but more often to local family farmers, who used them to formulate crosses to produce feeder pigs and butcher stock. The producers could assemble animals of desired breed, type, and conformation to create the hogs that would perform the best on their farms or for their customers. They crossed the purebreds to gain the fullest possible advantage of heterosis, or hybrid vigor, in the resulting offspring. Hybrids are stronger than their purebred parents; the result of two distinct bloodlines coming together is a stronger, more durable animal. But the traits that are expressed are less predictable for the same reason.

Now this is accomplished almost entirely by seedstock companies that sell crossbred and composite breeding animals. Hybrid hogs are really nothing new, and private and USDA research as early as the 1930s produced the various Minnesota, Montana, and Lucie hybrids of that era, which exist to this day. Many of the modern so-called hybrids are very complex in their structure, and because the strains are so closely managed, they lock producers into using a whole phalanx of company hogs to maintain even a semblance of hybrid vigor. It is a practice that could shut individual producers out of swine seedstock production in the same way that poultry producers were blocked from producing the now favored broiler and laying hen strains.

Seedstock and the Individual Producer

Purebred seedstock keeps production solidly in the hands of individual producers, affords them the maximum use of hybrid vigor, and maintains a sort of natural check on the price for this most important of inputs. To lose control of the seedstock pool is to lose control of the entire industry and the directions it might take in the future.

Talkin' Hog

Visit even briefly in the sale barn alleys, show barns, or hog lots and you will find the people there speaking a language uniquely their own. Not only will there be talk of gilts and barrows ("bars" in parts of the South and Midwest), but you will also hear about pigs with "daylight" that are "blown apart" or "coon footed."

Each new trend in swine type seems to add at least half a dozen new words to the swine raiser's vocabulary. Still, this is also a language with ancient roots. Where else would you find ancient words like *farrow* and *sow* used in the same sentence with *sono-ray?*

Farrow means to give birth. A *barrow* is a castrate, and a *gilt* is a female under 18 months of age that has yet to give birth. A *shoat* is a recently weaned pig. A *boar* is an intact male, and a *sow* is any female that has had a litter of pigs. *Stag* is a now little-used term for an older male to be sold for slaughter.

Older males and females retain good value in the slaughter trade for use in the manufacture of sausage and other highly spiced meat products. Whole-hog sausage is generally made with the pork from sows. However, given the current trend toward production in confinement, the heavier, higher-yielding sows are in limited supply, and most confinement-held sows are culled by or before their fourth parity and at a weight of 400 pounds or less. Heavy sows — 550 and above — now often bring as much or more per pound than No. 1 butcher hogs.

A *butcher* or *market hog* is one weighing 220 to 260 pounds and ready for sale for slaughter. These generally are five to seven months in age. A *feeder pig* is an animal generally weighing 40 to 70 pounds that is sold to a farmer/feeder to be fed out to market weight. Such pigs are generally between 8 and 12 weeks of age. They may be a bit heavier or lighter in weight. If they are lighter but healthy, they are young pigs and more prone to stress and setbacks. You have to wonder about their ability to hold up to the rigors of life in the finishing pen, and whether they are old enough to cope with marketing and transport.

If a feeder pig is much heavier than 70 pounds, it should raise suspicions that the animal is slow growing or stunted and overage for the weight. Traditionally, feeder pigs were 40-pound pigs between 8 and 10 weeks of age.

Every change in swine type — and by some accounts, there may have been as many as 20 in this century alone — produces several new descriptive terms such as those noted above. Short, thickly made hogs common early in this century were called *cob rollers;* later, taller, and flatter-muscled animals were called *race horses.* An animal that is *blown apart* demonstrates good *internal body dimension* throughout its entire length — the latter term refers to the visual indication of body capacity, internal organ development, and capacity to eat and grow. A good internal body dimension is associated with hardiness

Basic Hog Talk

Here are some of the key words that pertain to raising hogs. See if you can define them. If not, the definitions appear in the section above or in the glossary on page 285.

- Pig
- Hog
- Gilt
- Sow
- Shoat
- Barrow
- Boar
- Stag
- Farrow
- Butcher hog
- Feeder pig

and more efficient growth. Efficient growth, or *feed efficiency,* refers to how much feed it takes for growth: Good feed efficiency means it takes less feed to achieve good growth.

A pig with *daylight* has good leg length, because you can see some daylight underneath the animal. A *coon-footed* hog walks with a flatter and more flexible foot and has sloping pasterns that indicate it should stand up well on concrete or other unyielding surfaces. It has a "foot" that is, in many ways, akin to the foot of a raccoon.

Throughout this volume, I will use many of the terms in the modern swine raiser's lexicon. Some of them have been around for centuries, others for decades; some come and go in a year or two; one or two always seem to be working their way into the swine-raiser's vocabulary. A few years ago, the terms *spongy* and *Jell-O-middled* hogs came and went in popularity in little more than a matter of months. Such hogs had a thick middle when young; many believed this translated into a wider, thicker meat hog. Sometimes it did, but sometimes it resulted in a hog that was just fat. On the other hand, your great-grandfather would find assurance in knowing that a hog with a walnut on its side still has a small abscessed node — which is the same thing that *walnut* meant many years ago.

A feeder pig generally weighs 40 to 70 pounds and is between 8 and 12 weeks of age. An old rule of thumb is to buy colored pigs for cold or damp weather, because they will be hardier due to their colored-breed parents, and to buy white pigs for hot weather, because they supposedly have greater heat tolerance.

What really separates the pros from the tenderfeet is how the word *pig* is used. To be country correct, it is the term for a very young pig. A hog is a swine that weighs over 120 pounds. One farmer might tell another that the 400-pound boar standing before them is a "pretty good ol' pig," but the greenhorn who then strolls up and says, "Wow, look at that big pig," has shown himself to be totally green. You have to have earned your verbiage by time in the mud and the muck to be able to tell the difference between a "pig," a "hog," and a "shoat"; to know that a "piggy" female is in late-term pregnancy; and to recognize that a "he-boar" is not a comic book character (a "he-boar" is a pretty good ol' breedin' "pig").

Thinking Hogs

Far more important than talking like a hog raiser is thinking like one. What causes the success or failure of any farm enterprise is not the cost of doing business, disease, or weather, but the mindset of the producer.

If you could be guaranteed that hogs would always sell for at least 60 cents a pound and that corn would never sell for more than $2 a bushel, you still would not succeed if your heart and mind were drawn in other directions. To be successful with hogs, you honestly have to like them.

Cattle are prettier, sheep are cuter, and there are few endearing qualities about an animal with a battering ram on one end and a manure spreader on the other. Still, the hardiness and vigor of the hog has to be admired, and its earning power is legendary. As I told my new bride, Phyllis, many years ago, any time hog prices go over 45 cents a pound, the smell of hogs is the smell of money. And of course I must admit to a special fondness for the sight of a litter of baby pigs bedded down in bright, fresh straw.

Why Raise Hogs?

It might sound overly simple, but your first job is to determine why you wish to own hogs. When I was a young man, feeder pig production was the best way to get into farming, and many people started out with a handful of acres, a small bunch of gilts, and some portable, largely homemade equipment. Five gilts could become 50 in fairly short order, since they can be bred at eight months of age to farrow at one year. Within eight weeks of their birth, the producer has feeder pigs to sell. None of the other major livestock species can match this quick turnaround on investment.

There's nothing nicer than the sight of a litter of baby pigs bedded down in fresh straw.

Our first swine venture began with the purchase of a single Hamp-marked sow we named Esmerelda. At the risk of giving you another clue to my age, I will add that we paid all of $35 for her. Our purebred Duroc operation is closing in on its thirtieth year and we began it with the purchase of a single open gilt bought for $117.50 — and my share of the soybean check for that year was $120. Most of the great livestock herds and flocks in the world can trace their origins to just one or two females of merit.

Unfortunately or not, small- and medium-size farms can no longer stake their whole futures on just a single venture or two. Hogs, with their ability to breed and farrow year-round, however, can fit into a great many different plans of farm diversification. Three to five sows farrowing twice a year can produce handy-size groups of feeder pigs, herd replacements, or enough finishing hogs to fully utilize a 60-bushel self-feeder.

Identify Your Markets

Producers must be clear about the markets they wish to target with their hogs, the investment they can reasonably make, and their expectations for a porcine venture. State-fair winners have come from sow herds with five or fewer head, but those five were carefully selected and bred, and fed for optimum performance; they received their fair share of the producer's time and talents, and the producer was a true student of the art and science of swine raising.

Even the producer feeding out just one or two hogs each year for the family larder has to come to the task with some honest feeling toward the pigs and a plan of action for their purchase, care, and processing. A pig in the garden, if it gets there as a part of a plan, can do some free tillage and fertilization while on its way to becoming ham and chops. If it gets there through neglect or error, however, it can cost you the harvest of the garden and your religion until it is returned to its proper pen.

Hogs aren't for everyone, and they won't fit every farm. Still, with planning and knowledge, they can be made to fit more farms and country places than perhaps any other livestock species.

Two

Raising Hogs for the Family Table

MOST FOLKS START THINKING ABOUT hog ownership as a way to provide quality meat for the family table at moderate cost.

When you raise your own hogs, you select them, feed them to an exact slaughter weight, and direct the processing. Feeding out a hog for slaughter is "finishing" a hog. This gives you an assurance of quality and wholesomeness that you can have in no other way. And along with quality control, there is much that you as a home finisher can do to contain costs. Granted, a pig is not all chops, but when it is raised and processed to order you can expect to maximize the cuts and quality you and your family prefer in meat and meat products.

The Facilities

Where do you begin? Chances are you'll start with one or two feeder pigs. The first thing you'll need to think about is where to raise them.

Pig raising is simplest and most efficient in climates without great extremes in temperature or precipitation. The growout period will normally be between 90 and 120 days, depending on the starting weight of the pig or pigs, and can normally be fit into the spring or fall season to avoid the weather extremes of a Missouri summer or a Maine winter.

The finishing pen — the place where you bring your hogs to slaughter weight — should face south so that the house or sleeping area opens away from prevailing winds. Many hog owners place the pen at the foot of the family garden, where garden wastes can be thrown over the fence to the hog or hogs. Some even maintain two garden sites and pen a shoat or two in the fallow site, where the animals naturally till and fertilize the ground.

Small Finishing Units

I favor a small finishing unit — finishing being, again, the process of feeding out a hog in preparation for slaughter — with a capacity of two to four head; it can provide shelter in inclement weather, and it is easy to move if necessary. It keeps the hogs off the ground, which eliminates or reduces mud, and, in the process, also helps keep parasites in check. Basically, it is made up of two parts: a small house and a slatted, floored pen fronting it. Some call the latter half of this unit a "pig patio" or "sun porch."

Width

Width will be determined by the sleeping bed option chosen. Three-sided hog huts of sheet metal or sheet metal and wood construction can be bought in dimensions from 4.5 by 6 feet to 5 by 7 feet for about $130 new. In the Midwest, good used huts can often be bought for less than half of that figure. These huts are light enough and of the right size to form these feeding units. I have seen the all-wood 6- by 8-foot one-sow farrowing units placed on long runners where larger numbers of hogs are to be fed out. They will add to weight and cost, but can accommodate up to six or eight head.

The Foundation

Begin with two or three treated 4- by 6-inch runners of 12 to 16 feet in length. For a unit 4.5 to 5 feet wide, two runners should be adequate; for one that extends to 6 or 7 feet in width,

Insulate the Roof

You may choose to use plywood or sheet metal and 2-inch framing lumber to build sleeping beds. Just be sure to insulate the roof. This will prevent condensation — which will fall on the pigs and their bedding — from forming on the inside of the roof.

three will be needed. Elevate the runners at the corners and center points with two to four concrete blocks. The runners will form the foundation for the entire unit; elevating them on the blocks will facilitate cleaning the pen and make it easier to load and unload the hogs.

The Floor

Place a solid floor at least 2 inches thick on the half of the unit where the house or sleeping bed will be positioned. I prefer 2-inch full-cut native hardwood lumber such as white oak. It is durable and, unlike some treated wood products, will not be a skin irritant to lighter-colored hogs. Solid sheathing such as plywood may form a surface too slippery for hogs to stand and walk upon.

The "Pig Patio"

The outside pen or patio floor should be made from 2-inch-thick hardwood planking, with a 1-inch space between each board. The slots allow wastes to work through the floor and away from the hogs. At 1 inch, however, they will not catch the feet or legs of young pigs, nor encumber larger hogs. The pen's sides can be made of 1-inch-thick planking or 34-inch-high hog panels trimmed to fit the pen's sides and end.

Provide a 2-inch space between the bottom plank or panel bottom and pen floor to facilitate cleaning. In cold weather, the base of the unit can be shut off with bales of straw or sheet metal to prevent chilling drafts from coming up through the floor.

Advantages of a Small Unit

With normal care and maintenance, a small finishing unit can last 15 years or longer; can be moved about easily by a tractor equipped with a simple, two-pronged bale carrier; will not appear on property tax lists as a permanent structure; and will keep the hogs clean and easily accessible. This type of unit makes it possible to fit a couple of growing hogs into a space as small as 60 square feet while keeping down the mud and maintaining the hogs comfortably.

Managing Flies and Waste Breakdown

A couple of ducks can be an effective, natural method of fly control. Some hog owners also introduce red worms under their units to hasten the breakdown of wastes, for future use in the garden or wherever else composted wastes are needed.

Keeping Pigs on the Ground

Some owners prefer to keep their hogs on the ground. If this is your choice, you'll need to provide at least 150 square feet of pen space per shoat to keep mud problems from developing. In very wet or low areas, that space amount may have to be tripled.

Hog lots become muddy not from the rooting activity of the pigs, which can be controlled with the use of humane nose rings, but from excessive foot traffic and those sharp pointed hooves. In drylots, feeders and waterers should be placed on concrete pads or hardwood platforms to prevent mud holes from forming around such well-used sites.

You'll still need a house for shelter, such as the small unit described above, since it will be the animals' only dry retreat in wet or raw weather. It should have a step up to help keep muck out of the bedding area. Of course you'll eliminate the pig patio, since you'll be raising your hogs on the ground.

Houses in lots or pastures should still have flooring atop the runners or, if floorless, be placed atop a 4-inch-high pad or stone, packed earth, or cull lime. Blocking between the runner ends with planking and shallow trenching around the range houses will help to carry runoff water away from the hogs' sleeping area. In raw weather, a 4-inch-deep layer of clean straw will increase the comfort level of the sleeping area by at least 10°F.

Numerous variations on these housing themes are possible, but remember that your primary goals are to keep sleeping hogs dry, protect them from drafts, and get them up out of the muck. I have seen hogs housed in everything from junked cars to houses made from old pallets and shipping crates. They are doing just fine, thank you.

Fencing Options

To contain your growing hogs within a lot, you can use electric fencing, steel panels, or wooden gates. These fencing choices make it simpler to take down and rotate lots, or to enclose idle garden plots. With electric fencing, use two charged strands; one should be 4 inches above the ground, the other, 12 inches.

Hog panels are all metal, can be easily moved by one person, and attach quickly to steel posts. They are 34 inches high by 16 feet long. New, they sell for $15 to $20 each, but they can be bought used for about half that much. At one recent sale where very little in the way of other livestock equipment was offered, I bought five of these panels for just $20 total. To erect pens for growing/finishing hogs from such panels, 5-foot-long steel posts are adequate.

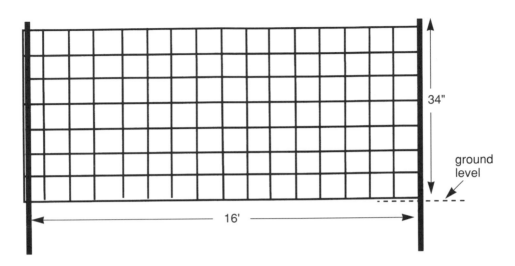

Hog panels are all metal, can be easily moved by one person, and attach quickly to steel posts.

I still see a few rolls of 26-inch-high woven wire around. If this is of heavy enough gauge, it can be rolled and unrolled repeatedly to make swine pens. Its height was originally set so that it could be unrolled between rows of field corn to contain hogs set there to glean or "hog down" the corn crop.

Essential Equipment

Besides facilities to house hogs and fencing to keep them in, essentials for hog raising include materials that will help you easily and safely handle hogs, even if you start with only one. The few simple pieces of equipment discussed below are all that you really need to comfortably and securely house one or more growing/finishing hogs. The feeling of comfort these provide will promote rapid and efficient hog growth along with ensuring the safety of both the animals and the producer.

Having your equipment in place before the hogs arrive allows you to avail yourself of any pig-buying opportunities that might arise. It will also help prevent you from rushing into expensive expenditures for equipment due to spur-of-the-moment pig buying.

Loading Chutes

A simple loading chute 22 to 30 inches wide and long enough to extend from the ground or pen floor to the truck bed can be made from little more than scrap lumber. Such chutes generally run from 3 to 9 feet in length. They should be kept narrow enough so that hogs cannot turn around inside them, but wide enough that someone can walk behind the hogs while loading.

Solid sides will make the hogs feel more secure in the chute; they will thus be easier to handle. Solid sides also prevent shadows, shifting light patterns, and outside activities from upsetting the hogs. Cleats in the floor will give them more secure footing. The chute decking should be made of material at least 2 inches thick. A few holes or small slots in the floor decking will allow rain and urine to run through, speed floor drying, and protect it from rot.

A bit of straw may encourage a hog to climb up an unfamiliar chute, but be sure to clean the chute floor after every use to extend its life. Inspect it often for raised hardware or broken boards that can cause injury to the hogs or the chute to fail. Drop a hog through a rotted chute floor and you may then have to carry it onto the truck to get it loaded and gone. Storing a chute indoors will certainly prolong its life, but if storage space is at a premium, it will suffice to give the chute a good cleaning after each use and set it out of the muck.

The best way to keep hogs moving up a chute is to walk behind them, blocking any retreat with a short gate or hurdle.

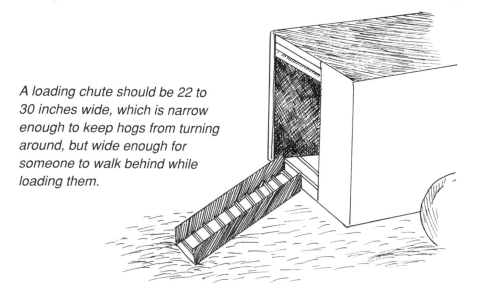

A loading chute should be 22 to 30 inches wide, which is narrow enough to keep hogs from turning around, but wide enough for someone to walk behind while loading them.

Handling Panels

Invest in an assortment of short, solidly made wooden gates and hurdles. Hurdle is an old term for a short, solid gate formed from 4- by 4-foot sheets of plywood. Hogs can't see through them, and they are useful for sorting, loading, and restraining the animals.

When snugly pressed between a barn wall or gate and one of these handling panels, a hog can be held safely for brief health treatments, sorting, or evaluation and measurement with a weight tape. These short gates are also useful for matching a loading chute mouth to various truck and trailer racks to provide safer and smoother loading and unloading.

You may have seen poles for handling hogs, but these are seldom used today. It's harder to direct hogs with poles than with gates or hurdles, and it's easier to become injured.

A hurdle is formed from 4- by 4-foot sheets of plywood. Hogs can't see through them, and they are useful for sorting, loading, and restraining.

Hands-On Handling

When handling young pigs and shoats of up to 70 pounds, there is no substitute for the hands-on approach. Handy "grab points" are the ears and hind legs. That tail makes an awfully tempting handle, but you should never attempt to lift a hog of any size by the tail. It is a part of the spinal system.

Nevertheless, as a last resort, I have tailed as well as eared many a shoat into the back of a pickup. And at weigh-in time for our local 4-H hog show, I've seen youngsters apply more grips and holds than I'm apt to encounter in watching a whole year of professional wrestling on TV. Over short distances, I will carry a pig by the hind leg and in a head-down position. If I carry it by the hind leg nearest me, the pig seems to travel more easily and more quietly.

Gripping Tongs and Snares

A snare or set of gripping tongs ($20 to $30 new) is also good for restraining hogs of market weight and even larger. The snare surrounds the hog's snout; the tongs apply pressure to the back of the neck. Both are pressure points and restrain the hog in such a way that you can administer any necessary health care treatments.

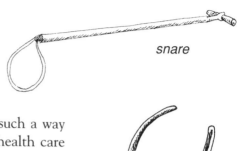

snare

ring pliers

A restraining device, either gripping tongs or a head gate, must be used when a ring is placed. Ring pliers clip a ring across the end of the nose or end of the gristly tissue around the nose. (See "Protecting Pastures" on page 128 for more on ringing.)

Feeding and Watering Equipment

Feeding and watering equipment can be as elaborate or as simple as your taste — or pocketbook — allows. The two essentials are to keep the drinking water clean and fresh, and to keep the feedstuffs palatable so none are wasted.

Homemade containers. Here at Willow Valley, we still use a lot of troughs made from old steel water heater tanks cut in half lengthwise. I weld bits of scrap iron to the ends to form legs for stability, and I weld bars across the tops to prevent the hogs from lying in them and thus wallowing out feed or water. We also use a number of feed and water pans made by sawing the ends off 30-gallon plastic barrels.

For a great many years we relied upon a homemade feeder of about 10-bushel capacity that was made with 2 x 4s, plywood, sheet metal, and a bit of found hardware. It was as ugly as homemade soap, but it wore like iron.

Store-bought containers. In the store-bought category, we have a number of rubberized pans and tubs and a couple of one-hole self-feeders suitable for one animal or a small group of 2 to 10 hogs. The pans and tubs are inexpensive and easy to move, and in cold weather they can be flipped over and stomped, which pops out ice.

A one- or two-hole self-feeder of sheet metal construction can be bought at farm supply stores for a bit under $100 and will hold 1 to 5 bushels of feed. We owned a few made from 2- by 6-inch lumber that lasted for many years when bolted solidly to a pen gate a few inches above the pen floor.

Fountain-style waterers. There are a number of fountain-type drinkers that can be attached to 55- and 30-gallon drums to provide drinking water in volume during warm weather. They can be set into a pen corner and hold enough drinking water for several days. Used, these fountains run $10 to $20 each; new, they are in the $50 to $60 range. They are durable and quite simple in design; I once owned one that I kept in service for more than 10 years.

A simple waterer that will work for one or two shoats and that will take up little in the way of pen space can be made from a 4- to 5-foot piece of 6- or 8-inch PVC (polyvinyl chloride) pipe, one pipe cap, and a nipple-type or Lix-it waterer head. Cap one end of the pipe, then affix the nipple through a hole low on that capped end. The nipple must be a gravity-flow model rather than a pressure type. When all of the gluing is thoroughly dry, the tube

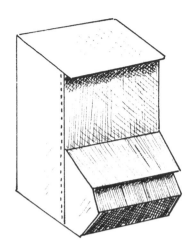

A one-hole self-feeder is suitable for one animal or a small group of hogs.

Countryside Publications

These young shoats are drinking from a freeze-proof automatic fountain. The balls move up and down, allowing the hogs to drink.

waterer can then be wired into a pen corner at the height appropriate for the pigs you are holding. The unit will hold a good amount of drinking water, is quite inexpensive, and can be easily stored when not in use.

Preventing feed and water waste. The flow on a self-feeder should be tightened down, so that a hog has to work a bit to get all the feed that it wants. Setting it to flow too freely can result in wasted feed. If you see feed on the ground around the feeder, you have a waste problem: As much as 10 percent of the feed you offer to your hog or hogs can be lost through simple waste. I place only a week's worth of feed into a self-feeder, to assure that it stays fresh and palatable.

> ## Essential Equipment for Raising Hogs
>
> ❏ Housing unit and pig patio; or
> ❏ Ground space and housing unit
> ❏ Fencing
> ❏ Loading chute
> ❏ Handling panels
> ❏ Snares or gripping tongs
> ❏ Waterer

If a hog is allowed all of the water it can drink in 20 minutes twice a day, its water needs will be met. Still, water is the most important of all of the feedstuffs. Troughs will keep water before the hogs around the clock in all but freezing weather. Bars or rebar rods welded across their tops will keep hogs from lying in the troughs and wallowing out the contents.

We have some deep wooden troughs put together with screws and caulked seams that we extend a short distance into our hog pens beneath a gate panel. Simple wooden legs support the portion of the trough that extends outside the pen. The hogs cannot wallow out any of the water, the troughs are easy to fill from outside the pen, and the drinking water in them stays much cleaner and more palatable.

Provide Adequate Trough Space

You should provide at least 12 inches of trough space for each hog in the pen to accommodate their herd instinct to eat all at the same time. One space or hole in a self-feeder can accommodate three to five head of growing/finishing hogs, because it keeps the feed before the hogs at all times. A water trough should be large enough to contain at least 12 hours' worth of drinking water.

Buying a Pig or Two

The old nursery rhyme goes: "To market, to market, to buy a fat pig, Home again, home again, rig-a-jig-jig." Well, buying a pig or two may not be quite as easy as this rhyme implies. Still, it is a fairly simple, straightforward task.

A meat hog of good type will yield about 85 percent of its liveweight in various cuts and meat products. Unfortunately, hogs are not all ham and chops, but one or two hogs fed out each year will go a long way toward meeting the protein needs of a typical family of four. If fed out at least two at a time the hogs will be more content, because they are herd animals. The second animal can be sold or used by other family members or friends. If sold, the money will help offset some of the costs of the animals and feed.

Where to Buy

The best place to buy feeder pigs is at their farm of origin. There you can often view the sire, dam, and siblings, which will give you an idea of what you can expect from the pigs as they grow out and reach finished weight.

Pigs at an auction are often stressed from the transport and handling. There is also a very real risk that they might have been exposed to disease organisms or sick pigs.

Countryside Publications

If fed out at least two at a time feeder pigs will be more content, because they are herd animals.

When to Buy

Early spring may be the most expensive time of the year to buy feeder shoats, because fewer sows are farrowed in the cold months of December, January, and February, when the early feeders are born. Still, these pigs are desirable, because they will reach a good slaughter weight before the weather grows excessively hot and humid.

Late summer through fall is also a good finishing period, as the pigs will usually be grown out before the late-autumn rains and winter cold begin. These are the classic pigs for the winter's meat.

Barrows or Gilts?

While they may grow a bit slower than barrows, I favor feeding out gilts. Not only do they hang a leaner carcass, but they can also be pushed harder with a hotter (higher-protein) ration. A "hot" ration might have a protein content of 15 to 16% or even 17%.

Barrows tend to pack on a bit more finish in the latter stages of the feeding period. In many parts of the world, intact male animals are fed out for slaughter to take advantage of naturally occurring male hormones, which seem to enhance growth rates. However, in those areas hogs are also slaughtered at a lighter weight and younger age, before any trace of a "boar taint" can develop.

Cost

Expect to pay a bit more per head for a single pig or two than the going rate for pigs in groups (droves). This is because the farmer needs to be compensated for the added bother and stress of working a whole group to sell just a couple of pigs. Also, he or she may find that the value of the remaining pigs is lessened by reducing the size of the group to be sold.

Most commercial finishers want to buy single-source pigs in groups as large as possible, to keep down handling stress. Also, the larger the group, the more they will generally bring per head.

A couple of rough rules of thumb: A 40-pound feeder pig should bring between 1.75 and 2.25 times the going per-pound rate for butcher hogs; a 40-pound pig should bring as much as the going price for 100 pounds of butcher hog. Add a bit more if you are going to buy just one or two head. Heavier pigs should bring correspondingly less per pound, and pigs down to 25 to 30 pounds a bit more per pound.

Tips for Buying Feeder Pigs

Here are a few tips to help you with feeder pig selection:

◆ The greatest variety and highest quality of pigs are available in the 40- to 60-pound weight range.
◆ Crossbred pigs are generally more vigorous and faster growing than purebreds.
◆ Pigs should be crosses from a planned breeding program and not simple mongrels.
◆ Buy pigs that have been castrated and healed if you are seeking barrows.
◆ Buy pigs that have been treated for both internal and external parasites.
◆ Buy pigs that are well past the stress and strain of weaning.
◆ Bear in mind that while gilts may grow more slowly than barrows, they will generally produce leaner carcasses and can be pushed harder with more nutrient-dense rations.

Prices will be higher when supplies of pigs are down and/or supplies of feed grains are high and the costs of finishing hogs are thus markedly increased. And high grain prices will, in turn, decrease feeder pig demand. Lightweight pigs sell for less in cold, raw weather, and more in the warmer seasons. Conversely, prices for heavyweight pigs trend upward in raw weather and times of high grain prices.

Evaluating Feeder Pigs

A good pig for feeding is not simply a scaled-down model of a finished hog. It should show indications of both good muscling and potential growth. Muscling is demonstrated by roundness or flaring in the areas where the primary pork cuts are to be found: the ham and the loin.

Growth potential can be seen in free and easy movement and a large, "stretchy" frame. Frame size is evident in the width between front and back legs, bone diameter as demonstrated in the legs, and width and depth of side.

In making final selections, many old hands focus on fine details such as width between the eyes, size of the jaw, or even foot size. Where these dimensions are of larger-than-normal size, the reasoning is that the rest of the animal will "grow to them." The width between the eyes is one I rely upon, as it does seem to be borne out by continued width throughout the body cavity.

When approached, a pen of pigs should rise up and move to the far side of the pen away from the intruder or intruders into their territory. They will then be lured back by their own curiosity to check out this new feature in their space. If they appear to act sluggish, they generally are ill or crippled.

Beginning the Selection Process

The pig-selection process should begin with your approach to the pigpen. Focus all of your senses on the pen as you close in on it. Listen for labored breathing, sneezing, coughing, or other sounds of respiratory distress. Certain diseases, such as TGE (transmissible gastroenteritis, an infectious gastrointestinal disease), create fecal material with a very distinctive odor. (More about TGE appears in chapter 9.) A cloud of dust or excess mud can indicate a stressful environment that might impair pig health or performance.

In selecting pigs, it is nearly always best to go with first impressions. Even the very best animals can be picked apart if you go over them hair by hair.

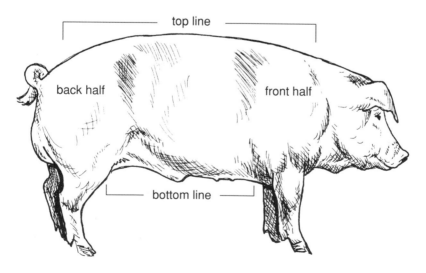

When checking out pigs, analyze them by parts: the feet and legs, the topline, the bottom line, the front half, and the back half.

Form a mental image of what you value and want to find in a pig, and compare the pigs driven before you to that image of the ideal pig.

For fastest analysis, break the live pigs into five overlapping zones: a) feet and legs, b) topline, c) bottom line, d) front half, and e) back half. I will go into more detail about these different areas a bit later, but they do help in finding that good pig that is the sum of good parts.

Note closely the environment in which the pigs have been raised, then try to buy animals from farm environments that most closely match what you have back home. This will do much to reduce stress on the newly arriving pig or pigs.

A lot of feeder animals are now being grown out in what are termed "controlled environment nurseries," which are temperature regulated and hold the pigs constantly indoors in small pens. Such pigs cannot always make the rapid adjustments needed for movement outside or to simpler, cold housing. Pigs are normally best moved from such units when they weigh 60 to 70 pounds; they should never be moved in times or seasons of extreme weather. In very cold weather they will pile up like cordwood and may even smother or crush each other or develop rectal prolapses.

Listen and watch closely as the pigs rise up and lie down. These are the times when feet, legs, and underlines are often the most viewable. Also, they are the times when the pigs are most often apt to cough, sneeze, or experience labored breathing due to respiratory ills.

Don't buy someone else's mistakes, accidents, or tailenders. Every pig has a history, and until you're satisfied that you know all there is to know about it, keep your hands at your sides and your wallet in your pocket. I have a friend who actually sits on his hands at most livestock auctions just to be sure.

Signs of a Healthy Pig

A healthy pig will display an alert and curious manner. Any pig you buy should also have these signs of vigor and good health:

- ◆ *A bright and clean haircoat.* To me, this is a must in feeder pig selection.
- ◆ *Free and easy movement.* It is vital if the pig is to remain sound and perform well while on feed.
- ◆ *Good growth for its age.* Simply selecting from the largest pigs in a pen or group can do much to build a selection program with an emphasis on individual performance.

Pigs to Avoid

Not buying someone else's problems simply means, "Don't buy any sick or sorry pigs" (see chapter 9). Here are some easily detected signs or indications of bad pig health that should immediately raise a red flag:

1. Pigs with exceptionally long and coarse haircoats and heads that appear too large for their bodies are generally overaged, stale, and badly stunted. "Unstunting" a pig is no simple task and is seldom cost effective.

2. Sniffling sounds as the pigs rise up and move about may be a symptom of respiratory ills, perhaps even atrophic rhinitis (see chapter 9).

3. Discharges from the nose and eyes or slight bleeding from the nose can indicate rhinitis.

4. Twisted, swollen, or misshaped snouts, are all indications of advanced atrophic rhinitis, a disease that robs greatly from growth and performance and that can lead to death. Do not buy from groups that contain these kinds of pigs.

5. Nodes or swellings (often called knots) in the jawline and other places are likely to be abscesses. They can be lanced and drained in some instances, but such procedures can introduce the abscess-causing organisms to your premises. Steer clear!

6. Dull, sunken eyes or a listless manner, are both indicators of an emerging health problem.

7. Dull haircoat, hair on end (a sign that the body is trying to trap heat), or an excessively dirty haircoat, are signs of stress. Pigs with hair on end and piled closely together are obviously chilled and undergoing severe stress. If you note a greasy appearance or a spotty haircoat, or an excessive amount of scratching, look behind the ears for lice or lice eggs attached to individual hairs.

8. Swelling in foot and leg joints can indicate injury or arthritis. Pigs presented in a ring or pen with little or no bedding generally have good feet and legs. Be on guard where straw bedding is piled deeply, however, as it can conceal much about foot and leg conditions.

9. Don't take a pig with a stiff gait, front legs that are too straight, hind legs that are tucked too far under the body, knocked knees, front legs that both "appear to be coming out of the same hole," or hooves with toe points of uneven size.

10. Dramatically obvious ills such as ruptures, sometimes called "busts," and broken or downed ears are reparable, but at considerable cost, and ruptures may be an inheritable trait.

Feeding Hogs

On the subject of swine feeding, I am reminded of an incident that happened a few years ago. On a warm fall afternoon, I sold a fellow about a dozen head of 40-pounders at a good price. After loading them, we sat down on the tailgate of his pickup for one of my favorite parts of hog production, the check exchange, and to shoot the breeze for a few minutes.

His first question was, "Now what do I feed them?" It was almost enough to cause me to let go of the check and unload the pigs. However, I decided to use his question as an opportunity to share some hard-earned experience on feeding hogs. It is information that I would like to share here, also. Above all else, remember that *the reason to feed hogs or any other livestock is not to save money, but to make money.*

The old rule of thumb was that in growing from 40 pounds to handy market weight (220 to 240 pounds), a hog would consume 10 bushels of corn and 125 pounds of protein supplement. The math still holds up pretty well, although many producers now feed rations that have higher crude protein levels or are more complex in their formulation.

Meat Scraps

In this century's early days of hog finishing, the primary protein sources were skim milk produced right on the farm and tankage or meat scraps. Tankage and meat scraps were and are by-products of commercial meatpacking and are dried and processed into a safe and easy-to-handle meal form. They run about 50 percent crude protein and, along with skim milk, fueled an old country belief that "It takes meat to make meat."

With the outbreak of "mad cow" disease in England, some question has been raised about the safety of feeding meat scraps. As best as I can determine, the real culprit in the British problem was brain tissue from sheep infected with scrapie. Such materials are not used in the manufacture of tankage or meat scraps in this country. Here, meat scraps are cooked products and contain no sheep parts. To me they remain very useful in booting up the protein levels in swine rations; they also continue the character a hog's diet would have in nature, including multiple animal protein sources.

Feeding Corn

Swine rations of that earlier time were simpler than most seen now — many hogs were finished on pasture or while gleaning grain fields. Still, as noted above, they were based on some nutrient-dense ingredients. Not only was the tankage supplement 8 to 10 percentage points richer in crude protein than today's widely used soybean-oil-meal-based protein supplements, but the corn in use then was also more nutritionally dense.

Those old, open-pollinated field corns often tested in at 13 to 16 percent crude protein. This was far better than the 8 to 9 percent protein levels of today's hybrid varieties, and it was nearly enough to serve as the sole component of late-stage finishing and maintenance rations. Add in the skim milk and red clover or ladino pasture and nutritionally — and maybe quite literally — it was hog heaven.

Yellow corn remains the basic feed grain for hogs and most other livestock species. Some form or variety is grown in every one of the 50 states; the national yearly crop is measured in billions of bushels, making it both widely available and modestly priced in most years; and it can be used in a number of forms for swine feeding. Many producers still feed some ear corn, sometimes

Feeding Alternatives to Corn

Alternatives to corn as the primary grain in swine rations include grain sorghum or milo, barley, and wheat. For best results, all should be ground to a meal form, and they should probably replace no more than 50 percent of the corn in the ration mix. The so-called bird-proof grain sorghums should not be used, because the factors that deter birds from feeding upon it as it ripens can also greatly reduce its palatability.

The newer yellow and white strains of grain sorghum are even better suited for use in swine rations, because they contain more vitamins than the red varieties of milo. Wheat may be an even better crude protein source than corn, but it is generally a more costly grain to feed; also, if ground too fine it becomes gummy when chewed, which can reduce overall feed consumption.

Barley and milo have about 90 to 95 percent of the feeding value of No. 2 yellow corn. Supplementation may be needed in the form of soybean oil meal or a premix pack. Premix packs vary in content and are generally used to add vitamins, minerals, and/or flavorings to boost nutrient density and increase consumption. For example, most rations containing milo are boosted with what is commonly termed an "A-D-E pack." It contains substantial amounts of the vitamins A, D, and E.

even allowing the hogs to glean it in the field, and others feed a great deal of shelled corn, especially if this feeding is on the ground or in open troughs to better counter waste.

A once-common practice was to keep barrels of shelled corn soaking in water to improve palatability and digestibility when fed. More common now is the practice of feeding corn in a meal form after running it through a hammer or burr mill of some sort. Some of these mills are also combined with a mixing tank that blends all of the ingredients of a ration to give it the bite-after-bite consistency that many now favor. Grinding the corn does improve its digestibility, and the hog is then better able to utilize the nutrients it contains.

Soybean Products

If yellow corn is the foremost feed grain in use, its counterpart in protein supplements is soybean oil meal, commonly abbreviated SBOM. It is a meal-type by-product of the manufacture of soybean oil. Varying in content from 44 to 48 percent crude protein, it is blended with corn or other feed grains, then supplemented with vitamins and minerals, to create a complete growing/finishing ration. It is also often the base protein supplement in young pig and breeding stock rations.

Soybean meal can also be found in two other forms. The first is commonly termed "hog forty" as it has a 40 percent crude protein content and is already supplemented with the more commonly required vitamins and minerals to form a complete swine ration. Mixed 4 parts grain to 1 part hog forty it creates a good growing/finishing ration with a 15 percent crude protein content. This 4:1 mix is known in the Midwest as the classic basic stock mix.

The other form is extruded soybean meal. This is made by running whole soybeans through an extrusion screw that processes them with both heat and pressure. Temperatures in the screw may rise to 300°F, and the resulting oily meal typically has a crude protein content of 36 to 38 percent. It does have several more percentage points of fat than regular soybean meal, however. This fat is a valuable energy source, increases ration palatability, and may even reduce some of the dust associated with many grind and mix rations.

There are alternatives to soybean oil meal and meat scraps, including some dairy-based products, but they tend to be rather higher in price. They contain more complex and easily digestible proteins. They do create very good, very palatable rations, and they are a must in formulating rations for very young pigs.

Using Dietary Additives

A number of dietary additives are offered for inclusion in swine rations with goals as varied as boosting nutrient levels, increasing consumption, and improving digestibility. More of us in the hog industry are now mixing in one of a number of regional products that contains fish by-products, yeasts, and kelp meal, to reduce performance setbacks during times of stress such as farrowing, weaning, and when hogs are moved. The one we happen to use at Willow Valley is Immuno-Boost, which is made in Texas.

In the past I have experimented with various probiotic products and top dressings for rations. Probiotics are starters and enzymes that activate or enhance the activity of the flora in the gut. Diarrhea can strip the gut of villi and "good" bugs (and so can antibiotics). Probiotics are supposed to help restore equilibrium. While some of these probiotic products have performed very well, I believe that fresh ingredients blended correctly are the real essentials of a good ration.

A practice favored by a number of old hands is to use at least 30 to 50 pounds of meat scraps in place of some of the soybean oil meal in a ton of swine feed. This gives the animals multiple sources of protein — just as they would obtain while foraging for themselves in the wild. A lot of these same veteran producers also grind and mix a bale of good alfalfa hay into every ton of complete feed they formulate. This is an added nutrient source and also gives the ration some extra bulk.

Choosing Types of Rations

From what to feed arises the question, What type of feed? That sounds almost too simple; just put what you want them to have in front of them and they will eat it. Right? Well, not really.

Early in this century, growing hogs were given X amount of corn each day. The corn was counted out to provide so many ears of corn per head or so many scoopfuls of corn per pen of hogs. For many months each year, the hogs were on legume-rich pastures with a bit of corn fed each day. Skim milk or tankage might be offered in a trough once or twice each day.

But swine feeding progressed. In its next step, hogs were offered access to corn — generally shelled corn — at all times, and offered protein supplements once a day in a trough or some sort of self-feeder. If all went well, the hogs were free to balance their own rations and grow at a good pace. The trouble was that the protein feeds were generally more palatable than the feed grain; they were consumed quickly and in ways that knocked rations completely out of balance. Not only did performance suffer, but hogs fell ill with gastric upsets. A better method of feeding hogs was found with the grind and mix method.

The grind and mix method. The next step in the progression of hog feeding was to grind the corn into a meal to increase its digestibility. (As much as 5 percent of the whole-grain corn fed to a hog can pass through its system totally undigested. The grinding process breaks down the kernels' seedcoats and makes the nutrients contained within far easier to digest.) The ground corn is then mixed with the other ration ingredients to form a "complete feed." These grind and mix rations remain in widespread use. They can be manufactured on the farm with a tractor-powered implement called a grinder/mixer, or a self-contained unit commonly called a stationary mill. Such rations can also be bought in bag or bulk form from many feed dealers and elevators.

The real plus with grind and mix rations is their bite-after-bite consistency of nutrient content and texture or form. With all in order, every pig that gets to the feeder gets a fully balanced, complete ration.

It is a very good approach to swine feeding as long as particle size does not become too fine. Very fine feed particles can cause problems of palatability and, in some hogs, can even trigger ulcers. Grinding and mixing also can add to feeding costs. If bought in bulk, however — there is a 1,000 to 2,000-pound (unbagged) minimum purchase at most elevators — it is one of the least costly forms in which to buy complete swine rations.

Pellets and cubes. The other commonly available form in which to buy swine rations is as pellets or cubes. Under heat and pressure and with a binder such as alfalfa the feedstuffs are formed into pellets or cubes that range from ⅜-inch long up to about the size of a man's thumb.

Pellets or cubes are a top feed choice for hog owners with small numbers of animals.

Julia Rubel

There are two real pluses with this process:

◆ It may free as much as 5 percent more of the nutrients contained within the feedstuffs (a result of the application of both heat and pressure to them).
◆ When fed in troughs or on the ground, pellets and cubes create less waste than meal-type rations. This is, however, the most costly form of ration processing.

Still, processing costs are quite a small consideration for the producer with but one or two hogs on feed. The ease with which pelleted feeds can be handled and fed, their widespread availability (in some areas they may be the only option for producers buying swine feed by the 50-pound bag), and the potential for increased efficiency make pellets or cubes a top feed choice for owners with small numbers of hogs.

Feeding Methods

There are two approaches to feeding out a pig or pigs to market weight. The first is termed "free choice" or "full feeding." Simply put, the hogs have access to all of the feed they might want 24 hours a day. By the numbers, a hog on full feed will eat about 3 percent of its liveweight daily in feedstuff. By the end of the feeding period, a finishing hog will be eating 7 pounds or a bit more of feed daily.

"Limit feeding" a hog to about 90 percent of appetite, on the other hand, will produce a slightly trimmer hog and will somewhat reduce daily feed costs. It will also extend an animal's time on feed; as a result, you will probably realize no overall savings on feed costs. Still, this is an option if you wish to produce some extra-trim pork for the family table.

As a finishing hog ages and grows, its growth rate and feed efficiency slow, and much of the weight gain in the late stages of the growout period is often finish (fat cover) rather than lean or muscle gain. A traditional feeding practice to help the producer cope with this natural pattern has been to reduce the crude protein content of the ration as the hog or hogs grow.

From 40 to 75 or even 125 pounds, the pig was kept on a ration with a 15 to 16 percent crude protein content, a true grower ration. The second stage — to about 175 to 200 pounds — was the grower/finisher period, in which rations with a 14 to 15 percent protein content were fed. From there to the desired ending weight, the true finishing period, many producers feed a ration with a crude protein content of a mere 12 to 13 percent.

Daily Consumption

To give you an idea of how much hogs eat, the following chart details the average daily consumption of our hogs on full feed.

Age	Weight	Daily Consumption	Feed
8 weeks	40 lbs.	1.2 lbs.	pelleted starter-grower 17–18% (crude protein)
12 weeks	75 lbs.	2.5–3 lbs.	grind and mix 11–15%
16 weeks	125 lbs.	4–5 lbs.	grind and mix 15%
To market	150–230 lbs.	6–7 lbs.	grind and mix 14–15%

Donald Bixby/American Livestock Breeds Conservancy

Limit feeding a hog to about 90 percent of appetite will produce a slightly trimmer hog and will somewhat reduce daily feed costs, but it will also extend an animal's time on feed. As a result, you will probably realize no overall savings on feed costs.

Today's leaner hogs, and especially the better gilts, can be pushed harder with richer feedstuffs. I keep my growing/finishing hogs on a 15 percent crude protein ration from 40 pounds straight through to end weight. I even use this ration with the boars that I grow out to 300 to 350 pounds before offering for sale. They continue to gain well and remain lean.

If you do opt to make changes in swine rations, do it gradually over a period of from three to seven days. Sudden ration changes can cause gastric upsets and stress in hogs of all ages.

Buying Feed

The producer with two hundred head of hogs on feed has a far greater incentive to shop for feed bargains than the producer with just one or two head, but some comparison shopping should still be in order. Few other livestock feeds have remained in the price parameters that have bracketed swine grower/finisher rations.

By the 50-pound bagful, feed regularly falls into the 8- to 12-cents-per-pound range; one growing hog will generally consume 650 to 750 pounds of such feed as it grows from 40 pounds to slaughter weight. Still, I have seen price differences as great as 1 or 2 cents per pound between comparable complete feeds at suppliers in our immediate area. I don't advocate switching brands every time you buy feed, but price savings like those matter even if you own just one or two head — enough to justify spending a bit of time on the phone with your local feed suppliers.

Selecting a Feed Dealer

There are more considerations in feed dealer selection than mere price. At some dealerships, small accounts are more appreciated and better served than at others. For example, when confronted with use and management questions, some feed dealerships can do little more than tell you to read the little paper tag sewn to one end of the feed sack.

I favor a small local elevator that seems to specialize in small accounts, carries a wide variety of feedstuffs, offers special services and delivery of amounts as small as 500 pounds, special orders products, and carries a good line of animal health products. While its prices on many products may be a bit higher, this is more than offset by its special services and considerations for the smaller producer.

Feed Storage

Storage of feed for one or two pigs need not be at all elaborate. Most producers buy feed from one week to the next. This allows them to keep their feedstuffs fresh, and to manage feed costs by averaging out price highs and lows throughout the entire feeding period.

Fifty pounds of feed can be dumped into a metal or plastic trash can with a tight-fitting lid. The feed will then be kept dry and protected from birds and vermin — the main causes of damage and contamination to feedstuffs.

If you do find a good price on feedstuffs and opt to stock up on them, a 55-gallon plastic or metal drum with a secure-fitting lid will hold up to 350 pounds of ground or pelleted feed or shelled corn. Set the barrel on concrete blocks or an old pallet to keep its bottom dry, and place a rodent control station safely beneath it.

A metal or plastic trash can is a good way to store small amounts of feed. It should have a tight-fitting lid.

Protecting Feed

Feed for hogs must be protected from mycotoxins, which are produced by fungi. These can accumulate undetected in swine feed, resulting in all sorts of problems ranging from impaired growth to a susceptibility to infectious disease. The USDA Animal and Plant Health Inspection Service (APHIS) cautions that young pigs especially are susceptible to the mycotoxins called aflatoxins, produced by aspergillus molds. Aflatoxins are often associated with drought-stressed corn and moisture levels of more than 14 percent. Fumonisins are another type of mycotoxin; at high levels they can cause rapid accumulation of fluid in the lungs.

Some mycotoxin problems are things a producer can't do anything about, APHIS says. What you can do to protect your pigs, however, is place grains in dry storage and keep moisture down. Clean feed storage bins often and use ground feed rapidly, so that mycotoxins don't have the chance to develop at threatening levels.

For even larger amounts of feed, I use a pair of old chest-type freezers in which the compressor or motor has failed. These are often available for the hauling from repair shops and appliance dealerships, and some of the larger models will hold up to 1,000 pounds of feedstuffs. The latching mechanisms must be removed as a child safety measure, but the door will still seal in place. These freezers can serve as quite satisfactory, watertight mini bins. They protect the contents from moisture and vermin, and they can be positioned all around a farmstead, even at penside. If you set them on a single tier of 8-inch-high concrete blocks, you'll find that they are both further moisture-proofed and their contents more accessible.

Bringing Your Pigs Home

With feed and facilities ready at home and pigs selected, the aspiring hog raiser is ready for the hands-on side of pork production. And it begins with getting the animals safely home and off to a good start.

If you're hauling them in a pickup or stake-sided trailer, cover the front of the racks with a tarp or sheet of plywood, to keep chilling drafts off the pigs while in transit. Bed the racks with straw to a depth of 4 inches in damp or cold weather. In hot weather, haul early in the morning or late in the day; use damp sand or sawdust for bedding.

At home, unload the pigs into a well-bedded sleeping area. It may be necessary to block the small animals away from a portion of the sleeping area, to discourage them from developing the habit of dunging inside the house. Wetting down a far corner of the pen in the first couple of days may encourage them to begin dunging in that area.

Using Self-Feeders

If your pigs will be allowed free access to self-feeders, tie the feeder lids up or open for a few days, until you are sure that the pigs are large enough and able to operate them. The same holds true if there are lids or covers on the drinking fountains.

Managing Stress

Moving pigs stresses them, and one sign of stress is going off the feed. It is not at all uncommon for newly moved pigs to go off their feed within a few hours. This shouldn't last for more than several days to a week, however, and the pigs should continue to drink water even while off their feed.

To help animals through this stressful time, many producers add a vitamin/electrolyte product to their drinking water. This helps the pigs maintain both health and body condition should they go off feed or have trouble adjusting to a new ration. Adding flavored gelatin to drinking water will further increase water consumption, and can also be used to mask any unpleasant tastes from medications that are administered through the water. (More on medications appears in chapter 9.)

Top-dressing feed with the same gelatin powder will draw the pigs to it and increase consumption there, too. Still, never forget that abundant drinking water is hogs' most important feedstuff.

Young, newly moved, and stressed pigs should not be offered anything in the way of new or exotic feedstuffs. Many producers try to buy some of the feed the pigs were previously eating or match it as closely as possible, to further help them through those first, stressful days at their new location.

Parasite Control

Unless you have been assured otherwise — and have some sort of documentation — assume that all pigs have at least been exposed to all of the major internal parasites or worms, as well as to external parasites and any localized problem parasites. They will need to be wormed. (Refer to chapter 9 for information on this.)

Hogs Aren't Garbage Disposals!

While on the subject of exotic feedstuffs, I should point out that hogs aren't four-legged garbage disposals. You can build pretty good hog rations from items as diverse as stale bakery goods and cannery wastes, but note that some very rigid laws now govern the processing and feeding of modern food wastes. Most states, for instance, require that food wastes be cooked to 180°F in an approved facility before they can be given to livestock. These laws apply even to the farmer raising one or two hogs for the family table, although they may seldom be enforced on small farms.

To build a truly balanced ration from ever-changing table scraps and garden wastes is nearly impossible. It is best to consider these extras, something over and above a pig's regular ration, and to never feed them in amounts so great as to disrupt the animal's digestive system.

Swine Finishing

Now that your hogs are settled in and you've got parasites in check, the tasks remaining in hog finishing may seem to amount to little more than keeping the feeder and waterer filled, but there's more — a good deal more.

Avoid Unnecessary Drugs

There seems to be an injection or in-feed antibiotic for every sneeze, sniffle, or hiccup a hog may have. When you are finishing hogs, however, your first job is to avoid the temptation to become "drug and medication happy." If a finishing hog has been carefully selected, is well fed, and is kept dry and comfortable, it should easily cope with most changes in the weather and shifts in the season.

There are swine feeds available that do not contain any type of antibiotic, although at times it may seem difficult to find them. We have found that hogs *not* on a steady diet of antibiotics grow as well as any others, and that if a problem does emerge they respond better and more quickly to a treatment-level dosage of a health product than those hogs given a steady low-level or subtherapeutic diet of the in-feed version of the antibiotic.

Signs of Health Trouble

Read chapter 9 for information on specific diseases. Suffice it to say here, however, that pigs that are slow getting off their beds, that are eating less, that stand up looking drawn, or that demonstrate any of the other signs of ill health need further examination and possibly treatment. They may have a bacterial problem or be coming down with an infectious disease. A rectal thermometer can also do much to help you diagnose health problems as they emerge. The normal body temperature of a hog is 101 to 102°F; a fever indicates that the body's systems are rising to its defense; whereas a below-normal temperature may indicate that body systems are shutting down, as occurs with kidney failure.

Note any health problems as soon as they start and summon the veterinarian as quickly as possible. With nearly any health problem, treatment is quickest and least costly if begun in the earliest stages.

If you have just one or two pigs on hand perhaps the best health appliance in which to invest is some means of restraint for the animal. This will allow the veterinarian to do his or her work safely and quickly.

Beware of Hot Weather

With growing and finishing hogs, we have encountered more health problems in hot weather than in cold. The first hot spell of the summer, especially if it comes on quickly and is accompanied by high humidity, can take quite a toll on finishing hogs.

In such conditions, hogs have had no time to become acclimated to the weather and thus react poorly to it. When it's hot out, provide extra drinking water and keep it fresh and clean. You may even wish to sprinkle hogs with water from above in the hottest part of the day. These measures will add greatly to hog comfort and survival.

(More on home medical supplies to keep on hand, as well as how to administer injections and treat wounds, appears in chapter 9.)

Finishing Weight

The last thing you'll have to determine when finishing hogs is what weight you want to feed them to. We personally prefer to feed our hogs for the freezer to just 220 to 230 pounds, or even a bit less, but we are just a two-person family.

Today's hogs can be taken to 250 to 260 pounds and still have quite lean and high-yielding carcasses. I have even seen some 300-pounders place quite well at market hog shows — although at that weight, we're talking about a lot of pork and some large primal cuts, such as the hams. Most families no longer really want or can use whole hams.

A hog will normally yield about 70 percent of its liveweight in pork and pork products. We generally feed out two hogs a year for our needs: one for slaughter in late winter and another for the fall. We also find fresh and cured pork to be a most appreciated gift item during the holiday season, so we do a bit of our Christmas shopping from the hog pen.

With your hog fed out to the desired weight, your next step is the harvest, or slaughtering process. In the next chapter I will discuss the various ways that you — the hog raiser — can become involved.

THREE

HOME BUTCHERING AND PORK PROCESSING

SLAUGHTER AND BUTCHER ARE HARD WORDS for many people, and I will concede that they are processes some of you may not want to take on as do-it-yourself projects.

They are traditional farming activities and can result in measurable cash savings, but they are also time-consuming and call for some special skills. Here at Willow Valley, we have slaughtered and worked up as many as four hogs in a two-day period, but of late we find it quicker and simpler to have a butcher hog custom-processed.

Costs to have a hog custom-processed locally now run $10 to $20 for trucking, $12 for slaughtering, 26 cents per pound for processing (cutting up the carcass, wrapping the cuts, and fast freezing the meat), a small rendering fee for any lard, and 29 cents per pound for special processing (curing, making into 4-ounce patties, making into links, etc.). A typical hog costs us $55 to $75 to work up. If we take it to be processed on a Monday the fresh meat is ready by Friday or Saturday, and the cured meat a couple of weeks later.

Beginning the Process

In most of the descriptions I've read butchering day dawns clear and cold, the smell of wood smoke wafts through the air, and geese fill the sky. For us, however, such days dawn when the bottom of the freezer begins to show and the butcher hogs are at last big enough. In even further contradiction, our butchering day begins in the late afternoon, when we drive the selected hogs to a pen near a scaffolding (which we erected for hanging the animals during skinning and cooling). There they are held off feed overnight.

The next day, family and friends gather at our farm for the butchering. We have at least one good shot on hand and a couple of strong backs to work under the experienced and watchful eye of my father-in-law, R. E. Perkins, Sr. To kill the butchers we use a .22 loaded with long rifle bullets; as a safety measure, only one person goes to the pen to kill the selected hog. For a killing shot, draw an X from ears to eyes, shoot just off center where the lines intersect. With the hog down, two of us quickly step in, grab up one front leg apiece, and drag the hog head first from the pen (this is with, rather than against the haircoat and is the best way to drag any downed animal). We then position the carcass with the head facing down a slope to facilitate blood flow and sever the jugular vein. Make this incision deep and long to be assured that it is free-flowing. Bleeding can take up to about 10 minutes; there will be a gurgling sound then reduction of flow before the process is complete.

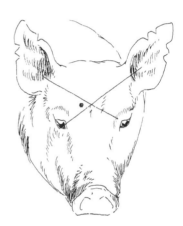

With the bleeding completed, we then move the carcass to the scaffolding, where we have affixed a pulley-type fence wire stretcher to the top beam. You'll find that a tractor with a front-end loader is useful in transporting, elevating, and suspending the carcass. Attached to the stretcher is a gambrel that we use to attach the hog to the stretcher and hold the carcass open while we work on it. For a gambrel we most often use a well-scrubbed single-tree from my father-in-law's collection of draft horse tack.

Imagine an X drawn from ears to eyes. Shoot just off center where the lines intersect.

Gentle Handling = Tender Pork

The less handling and stress a hog goes through just prior to slaughter, the more tender the pork will be. If a hog is fearful and becomes stressed, chemical changes take place that affect the meat. To reduce stress before slaughter, some farmers pen waiting hogs away from the slaughter area, so they cannot glimpse or smell any of the slaughtering or butchering process.

Skinning

Beneath the scaffolding's top beam is our main work area. Pick a site free from any obstructions, and one that has a safe surface upon which you can stand and work. We also work hard to keep this spot free from blood, wastes, and general clutter.

The hog is placed beneath the main beam, its back legs adjacent to the now fully extended fence stretcher. With a sharpened pocketknife we make a short, vertical cut inside each rear leg just above the hoof.

We then work carefully to expose the tendon that extends down the back of each leg. It appears as a white cord, very strong and taut, and even a slight nick will cause it to sever, so it must be exposed very carefully. We free each cord from the surrounding tissue, position the singletree between the hind legs, and then attach the hooks on each end of the singletree beneath the exposed tendons. The hitch ring on the singletree is hooked onto, or wired to, the lowered end of the fence stretcher. We can thus raise the carcass from the ground to a good working height. Should a tendon break or be severed, wrap heavy-gauge wire repeatedly around the foot and fasten it to a hook on one end of the singletree.

We begin skinning the carcass from circular cuts made just above each of the rear hooves. You'll see all sorts of fancy and high-priced skinning knives advertised, but I've found the old reliable three-bladed stockman's pocketknife to be more than adequate for this skinning task. We also keep at hand a sharpening tool or butcher's steel, along with a whetstone (dry hone).

With a bit of care two people can work at the skinning task — and this certainly makes the work go more quickly. From the cut ringing the foot, make another, much longer cut down the inside of each back leg, up to the vent and then completely around it. You can then remove the skin by working the knife blade between skin and muscle, pulling down on the skin as it is loosened. Another circular incision just through the skin at the tail head will enable you to completely skin the hams.

With the hams complete, make a long cut just through the skin from vent to head. Working from this incision, continue cutting, scraping, and working down the hide by pulling upon it as it is loosened. Special care is needed when skinning the flank area, because the skin is very thin there and must be worked back very carefully from the flank and belly.

As the work moves down the inverted carcass, the front legs are skinned out in a manner akin, but inverse, to the procedure for the back legs. Then make another cut completely around the head, and the hide should come away in one piece. At this point you should set the hide well away from the work area, to be disposed of following the dressing chore. It can certainly be processed for later hobby or craft work, but time always seems to be short at butchering time, so I generally dispose of the hide by burying it. It can also be composted in a large pile of sawdust positioned away from water sources.

You can now remove the head from the carcass with a series of twisting movements and careful cutting with a heavy-bladed knife or small hatchet. Set it aside for separate skinning and later use in distinctive pork products, such as headcheese. While a bit time-consuming, skinning the head is simply a matter of removing the skin from the back of the head forward, generally in modest-size segments. Take extra care when removing the skin around the jowls, where it is very thin, and when removing the fleshy end of the nose. The head can then be cooled out in a large container of freshly drawn water.

Kelly Klober

This is a carcass skinned halfway down. It can now be raised higher to give you easier access for the rest of the skinning process.

The Advantages of Skinning

Scalding is a more traditional process than skinning. (Instructions for it appear beginning on page 63.) Here at Willow Valley, however, we find skinning to be an easier and quicker process for our purposes. Skinning is also the safer of the two options. It eliminates the need to handle scalding hot water and the dangers of working around a scalding hot vat.

Evisceration

With the skin and head removed and the tail and all four feet in place, the carcass is ready for evisceration, or disembowelment. Starting again at the vent area, make a long downward cut to the head to begin the opening of the body cavity. Then, using a heavy-bladed knife or small hatchet, break open the pelvic girdle at the point where the hams naturally separate (often an assistant will aid with this task by pulling apart on the back legs). With this area opened, continue the cut the length of the body cavity, inserting the index and second fingers of your free hand into the cavity on each side of the knife blade to draw the skin away from the viscera while you make the long downward cut.

My father-in-law, R. E. Perkins Sr., splitting the sternum as part of the evisceration process.

Kelly Klober

You will also need the heavy-bladed knife to cut through the breastbone. With a barrow, the pizzle (penis) is dealt with at this time, too — simply cut it away and discard it. Should the urinary tract be full, tie it shut with a piece of clean baling twine, to keep urine from contaminating the carcass. A similar problem may be encountered with the bowel at the anus, and can be handled in the same manner.

At this point we place a large tub beneath the carcass to catch the viscera as they are removed from the body cavity. Then we begin the process of freeing the viscera from the anus and body cavity with the short blade of a pocketknife and the hands. All viscera should be handled carefully to prevent punctures, tears, or cuts, which could cause contamination of the carcass or organ meats. Much of the viscera can be removed in a single large mass, although the lungs, heart, and esophagus will have to be removed after the chest cavity is opened. We dispose of the viscera by burying them in a site

Tools for Skinning and Evisceration

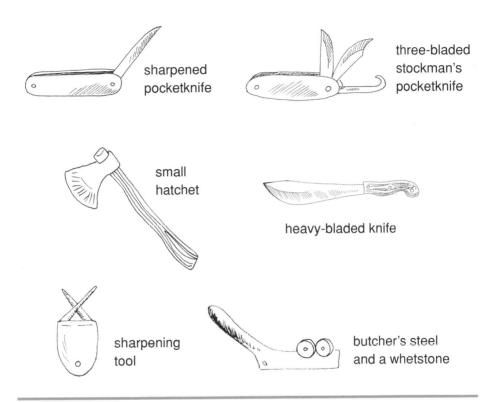

sharpened
pocketknife

three-bladed
stockman's
pocketknife

small
hatchet

heavy-bladed knife

sharpening
tool

butcher's steel
and a whetstone

well away from water sources, and to a depth that will discourage digging by wild or domestic animals.

Remove the liver carefully to protect both its quality and its future flavor. It is positioned adjacent to the gall bladder, and rough handling or a puncture could contaminate it with bilelike fluids. Split open the heart and rinse it free from all blood. We then place the organ meats in a basin of cool water, and put that basin in the refrigerator until it is time to package, label, and freeze the meat.

Remove the tail by making a cut ringing the tail head that extends through the back, then taking it out as you would a plug. Some producers scrape the tail and use it as a seasoning meat in certain vegetable dishes.

Cooling

With the viscera removed, we rinse out the body cavity with 10 to 15 gallons of cool water, or a bit more, and then leave the whole carcass to cool out overnight. Following tradition, many producers leave the butchering task for the coldest weather to hasten the cooling out of the carcass; a temperature drop of about 20°F from day into night, however, is adequate for the process. We have done butchering fairly late into the Missouri spring, and as early as late September in the fall. Where insects may be a problem, the carcass can be wrapped in cheesecloth without impairing the cooling process.

Cooling can be facilitated by halving the carcass lengthwise down the backbone. I've always found a cooled carcass easier to halve, however. We've used a hand meat saw for the halving task, but prefer a regular carpenter's handsaw, as the blade has less flex. A well-cooled carcass is also much easier to cut up and process.

By-Products

By-products of butchering include blood, feet, offal (the waste parts), and hides. While commercial slaughterers are equipped to salvage all of these items, the home processor is unlikely to be so equipped or able to sell them.

Hog feet do figure prominently in a great many recipes; they were a personal favorite of my wife's grandmother. Still, feet and offal, especially from several head of hogs, will pile up in a hurry, cluttering the work area and making footing and movement there more difficult. I generally trundle them away to be buried in a safe place, or share them with neighbors who have the time to process them correctly.

Blood is an ingredient in several types of sausages and other food items, but the time and effort required to gather blood and keep it sanitary can also take you away from the skinning of the more valuable carcass. Blood is very difficult to keep clean while being gathered, and unless several animals are slaughtered at once it is often difficult to gather in large enough amounts to be practical. Chitterlings, or small intestines, have the same handling and processing problems as blood, and their appeal is often quite limited.

Scalding

Although we prefer skinning, many hog producers still choose to dress hogs via the scalding method. It's the way some families have always dressed hogs, and they wish to continue the tradition.

Scalding takes more preparation time as well as more equipment and manpower than skinning. At least two strong people will be needed to move the hog or hogs in and out of the scalding bath. There is also the task of bringing the large amount of water needed to a boil and holding it there.

The scalding can be done in either a large (100-gallon) barrel or a scalding vat. The water must be heated to 150 to 155°F. Before scalding, the hog should be killed and bled as previously described.

To scald correctly, the hog must be completely immersed in the hot water. Two people using short log chains hooked about a lightweight carcass can lower it into a vat and manipulate it up and down in the water until the scald is complete. With a larger carcass, a boom or hoist is in order.

Before you remove the hog from the scalding water, make sure the scald is complete by checking to see that hair can easily be pulled away from each side of the carcass. To keep the scalding water hot when more than one hog is being dressed, many folks heat pieces of scrap metal in the scalding tank fire and then drop them into the scalding water.

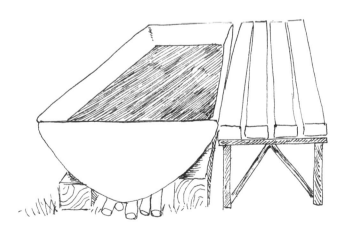

It will take two strong people using short log chains to lower a lightweight carcass into a vat and manipulate it up and down in the water until the scald is complete. With a larger carcass, a boom or hoist will be needed.

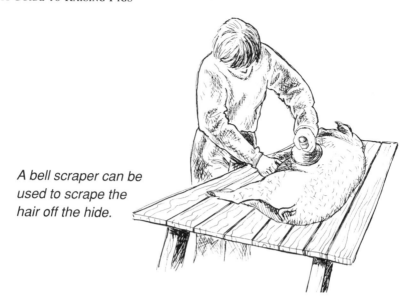

A bell scraper can be used to scrape the hair off the hide.

When the scalding is finished, move the hog from the water to a solidly constructed table for scraping. To remove the hair and scurf from the carcass, begin by scraping the legs, head, and belly. Keep the carcass wet with warm water throughout the process. When you have completed the scraping, you can handle the carcass just as described for skinning.

Breaking Down the Carcass

Meat cutting is both a skill and a science. As a home processor, you certainly do not have to be as precise in your work as a retail butcher when reducing a carcass to its most usable-size portions. This often involves little more than following the natural seams within the muscling.

Pork is made up of muscle fibers, connective tissues, fat, juices, and water. Young animals will have more tender muscling than older ones; physical activity also affects muscle quality. The most tender cuts come from the loin region — the boneless cut from the pork loin is, in fact, called the tenderloin. Tougher cuts of meat have thicker muscle fibers and more connective tissues. They come from the muscles the animal uses the most.

Hogs have both a fat cover over the carcass and fat flecks within the meat, which make the meat tender, juicy, and flavorful when cooked. Without a certain amount of fat, pork or any other red meat would be largely unpalatable. Among swine breeds, the Duroc and Berkshire are most noted for the flavor and eating qualities of their pork, even when they are crossed. Fat flecks within the muscling in these breeds function in much the same way that marbling does in good beef. The Japanese are now especially partial to the pork from Berkshires and their first-generation crosses, and such pork brings a premium when sold in the Far East.

Immediately after cutting, meat has a distinctive purplish red color, but exposure to oxygen soon turns it to the bright red "meat" color with which we are most familiar. To break down a hog carcass, we use only a long-bladed boning knife, a heavy-spined butcher knife, and a meat-cutter's handsaw. We send out our fresh hams and shoulders to have them sliced to the desired ⅝-inch thickness on a power bandsaw for a small fee. For such slicing, partially frozen pork is easier and safer to work with than fresh pork at air temperature.

Primal Cuts

Primal cuts are large cuts that are often transported to butcher shops for further butchering and sale. There are seven primal cuts in the halved hog carcass:

1. The **leg**, which is comparable to the round in beef and produces boneless leg, ham, and ham slices or steaks
2. The **loin**, which can produce blade chops, loin chops, butterfly chops, country-style ribs, back ribs, Canadian-style bacon, loin roasts, and tenderloin
3. The **side pork**, which yields the bacons
4. The **spareribs**, which yield both ribs and salt pork
5. The **Boston shoulder**, from which can come pork cubes, Boston roasts, shoulder roll, and that midwestern favorite, pork steaks
6. The **picnic shoulder**, which yields up roasts and steaks, ground pork, and sausage
7. The **jowl**, which can be cured for seasoning meat or sliced like bacon

No single carcass can produce all of the above, but the beauty of home processing is that you can give over as much of the carcass as possible to your family's favorite cuts.

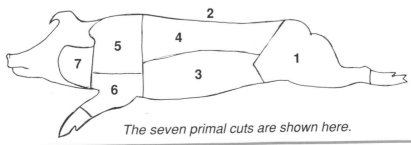

The seven primal cuts are shown here.

Butchering

As noted earlier, the butchering process is essentially the disassembling of the carcass at the joints and along naturally occurring seams within the muscling. Having the right equipment will help you obtain the cuts of meat you want.

Butchering Tables

I have seen butchering done on the kitchen table — and done a bit of it myself — but a more stable, easy-to-clean work surface is preferable. Workbench-type tables topped with a fiberglass laminate, or even stainless-steel restaurant tables bought secondhand, are excellent for supporting butcher work.

One of midwestern furniture makers' earliest uses of walnut was in simple butchering tables. It is a sturdy wood that stands up well to the oils and greases common in the butchering process. Valuable though walnut furniture may be now, here in hog and corn country it was used initially for this most utilitarian of purposes.

A 10-foot-square area in a shop or outbuilding with smooth and easily washable floor, ceiling, and wall surfaces is the ideal site for a home butcher shop. It should be centered on a large worktable that can be accessed from both sides and both ends.

Cutting Tools

Quality tools kept sharp and clean do much to expedite and ease the butchering task. Cutlery for breaking down the primal cuts into table-ready portions includes a boning knife, butcher's knife, cleaver, and meat saw (a short handsaw will do in a pinch). A butcher's steel and dry hone will also be needed to keep a good working edge on the cutlery. Dull edges slow your work and increase the risk of injury. Cutting through hair or bone seems to take an added toll on cutlery edges.

You will need a small meat grinder for turning scraps and trim into sausage or pork burger. A bandsaw with a meat-cutting blade will certainly simplify the process of slicing steaks and chops into the desired serving thicknesses.

Separating the Carcass

Separate the carcass into the larger, primal cuts with the butcher's knife (an 8- to 10-inch, heavy-spined blade); use the cleaver where necessary to

Meat-Cutting Tools

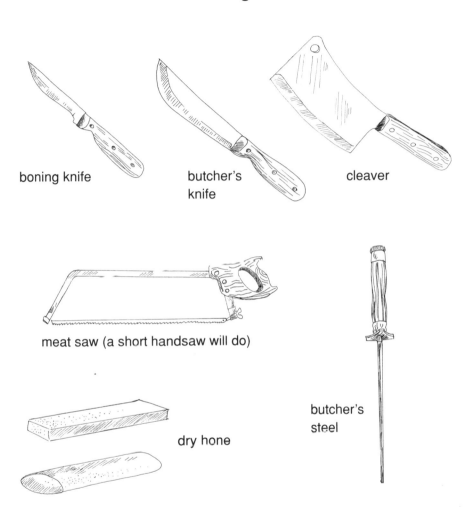

boning knife

butcher's knife

cleaver

meat saw (a short handsaw will do)

butcher's steel

dry hone

disjoint major carcass segments. Cutting completely around the joint points and sharply rotating the primal cuts to be detached will often hasten the joint's separation. As a safety measure during this work, wipe the cutlery handles and your hands often to ensure a solid grip on the tools at all times. There are gloves you can buy that have material on the palms to give you an especially good grip.

With the primal cuts separated, you must then decide how to further break them down for table use and long-term storage. For example, you can debone the loin to yield the tenderloin, which can then be sliced into medal-lions or butterflies, or you can leave the bone in and slice the loin into pork chops. The ham can be left whole, broken down into two or three large roasts, or sliced into ham steaks of various thicknesses. Fresh, uncured ham is often said to be "green."

Chilled meat is the easiest to work with; some folks even partially freeze pork that is to be ground into sausage or burger. This is a virtual must if the grinding is to be done with a small, hand-turned grinder. Steaks and chops are normally sliced to a thickness of ⁵⁄₁₆ inch for fast and even cooking. Roasts are normally broken down into segments weighing 2 to 4 pounds — large enough to center a meal for the entire family.

Perhaps at this point I should repeat that the costs of butchering, and your own personal tastes, may favor having your hog or hogs commercially processed. Commercial operators have the skinning and cutting tools, refrig-eration for rapid chilling, carcass-handling implements, and manpower to do the job at a fraction of the toll it takes on many country households. When the butchering begins, virtually all effort must be given over to it until it is completed.

Think of it in these terms: A 210-pound butcher hog (quite small by today's standards) will still have a dressed weight of about 150 pounds and a packaged or wrapped weight in the area of 120 pounds. At home, those 75-pound (or heavier) carcass halves must be carried to the work area and handled largely by just one person. The meat must be kept clean and wholesome and moved to proper long-term storage as rapidly as possible.

Clearly, butchering even a single hog can fill up the family schedule for the better part of two days, and once started it is a task from which you cannot withdraw.

On the other hand, even if you wish to add some personal touches to your meat, you can do so with commercial processing: Simply request the return of a green ham or bacon from the processor for home curing.

Putting Away the Pork

When you're up to your elbows in the task of working up a carcass, you'll soon see the truth in the words of the legendary livestock nutritionist Frank B. Morrison in the ninth edition of his classic text, *Feeds and Feeding Abridged*: "Pigs excel all other farm animals in the efficiency with which they convert feed into edible meat. They require much less feed and much less total digestible nutrients for each pound of gain in live-weight than do other farm animals. They also yield a higher percentage of dressed carcass, a larger percentage of the carcass is edible, and pork is higher in energy content than other meat."

The half of the aforementioned hog with its 75-pound hanging weight should yield 14 pounds of fresh ham, 12 pounds of loin or chops, 6 pounds of trim for sausage, a 12-pound bacon, 3 pounds of spareribs, a variable amount of lard (up to about 10 pounds), 11 pounds of shoulder for steaks or roasts, and 5 pounds of bone and shrink. Many of these weights include the bone, however, and thus are not 100 percent edible meat. Still, there is very little waste with modern hogs. I well recall sending a gilt to be processed and actually having to buy a few pounds of fat to add to the trim to make a sausage that would have enough fat for proper cooking.

What, then, are you to do with all of this protein largesse? How can you put it away for a rainy day — or a B.L.T. on an August afternoon when those tomatoes are truly ruby ripe and fresh from the garden?

Pork products can be frozen, canned, cured, and/or smoked. Freezing is now the most common means of home meat storage, and it is also perhaps the simplest and least time-consuming.

Freezing

The deep freeze or freezer is not a wonder appliance into which you simply drop things and find that they remain fresh and appealing for forever and a day. Understanding a freezer's limits will prevent both potential illness and substantial financial loss.

How to Preserve Quality

To preserve their quality, you should freeze pork and other meats as quickly as possible. Set the freezer control to 0°F or slightly below; higher freezing temperatures will cause larger ice crystals to form in the meat. Those crystals break down meat cells and fibers and adversely affect juiciness and texture in the meat when it is then prepared for the table. Rapid freezing, on the other hand, causes smaller ice crystals to form in the meat and thus preserves better eating quality. When the meat is completely frozen through, you can return your freezer to a higher temperature setting.

Freezing simply slows the changes that affect food quality. Bacteria are not killed by freezing; they are simply halted from multiplying. How the meat is prepared for freezing, how the freezer is maintained, and how foods from the freezer are thawed prior to preparation will all affect eating quality and wholesomeness. Freezer management is truly an ongoing chore.

One cubic foot of freezer space holds approximately 35 to 40 pounds of food, and 4 to 6 cubic feet of freezer space should be budgeted for each family member. A well-filled freezer operates better and more efficiently than one that is only partially filled.

You should attempt to freeze only those amounts of meat that will freeze thoroughly in 24 hours. Also, never try to freeze more than $\frac{1}{15}$ of the freezer's capacity at a time. A good guideline is to freeze only 2 to 3 pounds of meat for each cubic foot of freezer space.

To save space in the freezer, debone the cuts of pork as much as possible, and trim them into pieces as uniform and compact as possible, before wrapping them and packing them into the freezer. Because frozen meat is best when consumed as soon as possible following thawing, it should be wrapped in portion-size amounts that can be used at one time or for one meal.

Tips for Freezing and Thawing Pork

While freezing hastens color changes in red meat, when pork is properly handled and wrapped, and then used within the approved storage times, its eye appeal, flavor, and juiciness are little affected. Do not, however, season ground pork before freezing, because most seasonings — especially sage — are intensified by the freezing process. Lard, after complete cooling, can also be stored in the home freezer. Freezing times that will best maintain the table qualities of various pork items are:

Cured bacon	1 month or less
Cured ham	1–2 months
Fresh pork chops	3–4 months
Pork roasts	4–8 months
Sausage	1–2 months
Organ meats	3–4 months

The best way to thaw frozen pork is in the refrigerator. For small roasts and steaks, allow three hours of thawing time per pound; for large roasts, four to five hours per pound.

After wrapping, label the freezer packages with a grease pencil, permanent ink, or a tape-type label. The label should note the cut contained within and the date that it went into the freezer. Those pieces of meat that have been in the freezer the longest should be placed nearest the door for soonest use.

Wrapping Pork for Freezing

All materials used for freezer wrapping should be both vapor and moisture proof. Heavyweight aluminum foil or freezer-grade plastic wrap is the best choice for wrapping bulky or irregularly shaped cuts. When you're packaging steaks, chops, or patties, place a double thickness of waxed paper between them for easier separation after they're frozen. Once you have properly labeled them, spread the new packages out in the freezer as much as possible to hasten their thorough freezing.

The basic closure technique for freezer wraps is the three-step drugstore fold, which forms an airtight seal that will protect the contents from freezer burn:

Step 1. Place the pork cut in the center of the wrapping material.
Step 2. Bring the two horizontal ends together and fold over until they are tight against the meat.
Step 3. Tightly fold one end and then the other, turn each end underneath, secure tightly with freezer tape, and label fully.

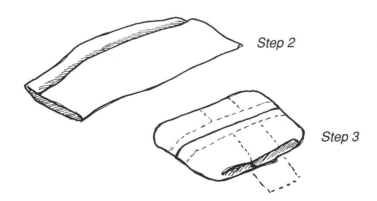

Fold wrapping material tightly against the meat, as shown, and secure with freezer tape.

Ground Pork

With the carcass of a hog broken down and cut up, you will have a good-size pile of lean trimmings for sausage. Before grinding, all animal heat must be gone from the meat; it should also be chilled, to firm it up. Cut the chilled meat into cubes or strips for easiest feeding into the grinder.

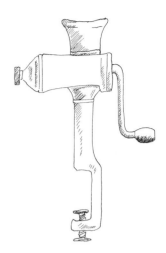

A metal hand-turned grinder is sufficient for processing small amounts meat.

An electric grinder will certainly speed up the pork-grinding chore, but a hand-turned grinder is adequate for small amounts of well-chilled or partially frozen trimmings. All-metal grinders offer the greatest durability and ease of cleaning. They will either bolt to a large board or table or clamp to the edge of the work surface. They come with grinding plates of different sizes that you can use to double-grind the meat to the desired texture.

For best taste, some fat should be ground with the lean meat to enhance the juiciness and flavor of the sausage. The fat content of pork sausage normally ranges from 20 to 30 percent; the latter is typical of many supermarket sausage products. We have some of our Willow Valley hogs custom-processed to produce an 80 percent lean whole-hog sausage, which we sell at $2 to $2.50 per pound at our local farmers' market. The whole-hog sausage sold in most supermarkets is generally made with pork from cull sows rather than from younger, pricier butcher hogs.

Pork for sausage should first be run through the coarse blade of the meat grinder. Then mix it thoroughly with your hands; spread it out thinly and season it evenly; mix it thoroughly again by hand; and regrind it through a finer plate (I recommend the ⅛-inch-hole plate). Sausage to be frozen is best left unseasoned, and sausage for canning should receive no sage seasoning.

Pork Burgers

One of the newer pork products is one commonly termed "pork burger." It is an 80 percent lean, ground-pork product seasoned with one of a number of various seasoning mixes. The seasoning is blended in at the rate of ½ ounce per pound of ground pork. Pork burger is very good on the barbecue grill, and it makes one of the richest tasting cheeseburgers I've ever had. It can be stored for up to four months in the home freezer.

Lard Rendering

Another by-product of hog processing is fat for lard making — but only if the hog has been scalded, because skinning removes too much fat. Lard is a useful cooking product and the secret ingredient in the crusts of more blue-ribbon-winning pies than you might imagine. With a kettle of bubbling lard and a stringer of fish, you'll soon have an old-fashioned fish fry on your hands.

The fish will take a bit of doing, but the steps for rendering lard are fairly simple. Here are the instructions, along with some tips.

1. Remove all of the skin and lean from the backfat and all other fat trimmings. Fat from around the internal organs will yield a darker lard; it should be rendered separately, since it is usually discarded.
2. Cut the fat into small but uniform-size pieces (about 1- to 2-inch cubes).
3. During the rendering the fat should remain at around 212°F. Never allow it to rise above 255°F, because scorching can occur. If you are rendering in a kettle over an open fire, keep the fire low and stir the rendering fat often. Do not render the lard in a copper or brass kettle; these metals can cause it to become rancid.
4. As rendering continues, cracklings will begin to rise to the surface and turn brown. Skim them off. Hot cracklings can be pressed in a lard press to produce more lard. Cracklings are also a pretty fair snack food if you were raised country. They can be added to your favorite corn bread recipe to make cracklin' bread, and they will keep your hounds fat and slick like nothing else.
5. When the boiling ceases and all the water in the lard is evaporated, the rendering process is complete, and the heating should be stopped.
6. Strain the lard through several layers or thicknesses of cheesecloth into pails, crocks, or other storage containers.
7. Cool the lard immediately at a temperature near freezing. (To set properly, it must chill through gradually.)
8. As the lard cools, stir frequently to achieve the desired creamy appearance. This will also prevent the lard from taking on a grainy texture. Dark-colored lard either was scorched or had too much lean remain attached to the fat.
9. Air and light are harmful to lard, so containers should be filled to the top, sealed tightly, and stored in a cool, dark place. Package lard for freezing in amounts that you can use quickly following thawing.

Preserving Pork

Drying, smoking, and curing are the oldest known methods for preserving meat. They yield a distinctive flavor, a taste of history some might say, and no wonder — even the great early Roman Cato was known to have had a favorite recipe for salting down hams. The practice of drying and then smoking meat can be traced all the way back to the ancient Egyptians and Sumerians.

Pork-preserving methods are time-consuming by today's standards and the finished products may require refrigeration for safest storage. Still, their taste, aroma, and texture can be produced in no other way.

Most cures are quite often closely guarded family secrets and may include such widely varied special ingredients as red pepper and cloves, but the two basics of most cures are *sugar* and non-iodized salt. Salt is both a good preservative and a flavor enhancer. It moves through the meat by that process we all read about way back in high school, osmosis. It also helps inhibit bacterial growth. Sugar helps counter some of the salt's harsh edge, brings out further flavor in the meat being cured, and lowers the pH of the curing solution.

The third ingredient common to many pork cures is *saltpeter* (nitrate), which is different from curing salt. It fixes the desired taste and color in cured meat and protects it from the often fatal botulism organisms. The nitrate question is a touchy one for many. Nitrate has been associated with some forms of cancer, but is the best protection the home processor has against botulism. If used, saltpeter should be stored and handled with great care and added only in the exact amounts set down in the various curing recipes and formulas. Some people delete the saltpeter, but use the cured product quickly.

The best pork for curing comes from young slaughter hogs in the 180- to 240-pound weight range. Meat that has been frozen and thawed should never be used for curing, however. Ice crystals damage meat texture, so meat that has been frozen will be more vulnerable to spoilage. Hams for curing should be from hogs that have been scalded rather than dressed out by the skinning process; because the skin (rind) is necessary to maintain juiciness and texture. I have no way to substantiate this, but it is my belief that the pork from hogs fed whole-grain corn and grown out more slowly has better table and curing qualities.

Quality Assurance at Home

For the assurance of wholesomeness and quality, home processing of pork and pork products cannot be surpassed. This task may at first appear to be beyond your skills or desires, and allowing a custom processor to do some or all of the pork-processing tasks is certainly a reasonable and economically viable alternative. Still, the savings, assurance of quality, and freshness and taste of home rearing of a meat hog or two do make this a most worthwhile project even if you have just a small bit of acreage.

Curing Hams

The two most common types of cure are the *dry cure* and the *pickle cure*. In the dry cure, ingredients are rubbed into the surfaces of the meat. With the pickle cure, a brine-type solution is injected into the meat down to the bone with a large needle and syringe.

Dry-Cured/Sweet Pickle-Cured Ham

This combination cure is from the University of Missouri. A pumped pickle cure is included because the quicker the cure reaches the center of the ham, the less risk there is of meat spoilage. A needle and syringe, which can be purchased from a store that sells cooking supplies, is the recommended way to apply the pickle cure.

7 pounds meat-curing salt
3 pounds white or brown sugar
1 gallon water
1 5-inch-long needle attached to a plunger-type syringe
  (available from many cooking utensil suppliers)

1. To prepare the pickle cure, dissolve 2 pounds of the salt and sugar mixture in the gallon of water. Pump no more than a volume of the pickle cure solution that is equal to 10 percent of the ham's weight.
2. Rub the remaining dry cure into the surface of the ham at the rate of ½ ounce to 1 pound of ham.
3. Allow the ham to cure on a table at 34–45°F for 14 days.

The term *country cured* is enough in and of itself to set most mouths to watering in anticipation, and few delicacies are more eagerly sought out than true country-cured hams. Here in Missouri a number of family farmers have gone into the business of producing country-cured hams and now ship them worldwide as gourmet treats. One of the most competitive events at many county fairs, as well as the summer-climaxing Missouri State Fair, is the cured-ham competition.

Generally, the "country-curing method" is begun in December or January when the cold weather has settled in for a good long spell. You'll want to begin with hams chilled to 40°F that come from hogs slaughtered within the last 24 to 30 hours; this cure must be applied within 48 hours of slaughter. We are talking truly fresh here, and one of the high points of what is traditionally a week or more of butchering and related activities.

Few delicacies are more eagerly sought out than true country-cured hams.

Country-Cured Ham

Trim your ham of excess fat, to give it the traditional ham shape, while being careful to expose as little of its lean portion as possible.

> 2 pounds non-iodized salt
> 1 pound white or brown sugar
> 1 ounce saltpeter (optional here)

1. Mix the ingredients together thoroughly. To further enhance flavor and aroma, you can add black pepper, red pepper, or cloves to taste. This recipe should provide enough curing mixture for 2 hams, the production of one hog.
2. Rub the cure into all surfaces of each ham at the rate of 1¼ ounces of cure per pound of ham.
3. In the single-application method, all of the cure is applied to the meat at once, with the lean areas receiving most of the mixture. See that the hock receives plenty of the cure as well, and that it is likewise applied uniformly to the face of the ham.

4. Wrap the ham completely in nonwaxed paper, place in a cloth bag, and hang shank down in a well-ventilated, dry place. Allow 2½ days of curing time for each pound of ham, plus 1 extra day for each day the ham might have been frozen during the curing process.

5. When the curing time is up, unwrap the ham, remove the excess cure, rewrap the ham, set it to age in a cool (40°–50°F) place, and then check it weekly for signs of possible spoilage.

   In the two-application method, initially apply ¾ ounce of cure per pound of ham. Wrap the ham and place on a shelf in a cool, dry place for seven days. When that time is up, apply the remaining ¼ ounce of cure per pound of ham, then handle exactly as above.

The Missouri-Cured Ham

*From my father-in-law, R. E. Perkins Sr., comes this Missouri cure
for the hams from a 200-pound butcher hog — a weight that has to be considered
prime for the family table.*

    2 cups meat-curing salt
    1 cup brown sugar
    2 tablespoons black pepper
    1 teaspoon red pepper

1. Rub the mixture thoroughly into all surfaces of the ham.
2. Wrap the ham completely in newsprint.
3. Tie the wrapped ham with a string in a 1-inch-square, crisscross pattern.
4. Place the ham joint down in a muslin sack.
5. Hang it in a cool, dry place and let it drip. If the ham doesn't start to drip right away, take it down and examine it carefully for problems such as mold or insects.
6. Let the ham hang and cure for 6 to 8 weeks.
7. When the curing time is up, take the ham down and wash off the remaining cure. The ham can then be used right away, refrigerated, or frozen.

This is a very old recipe. If you are concerned about the inks and processing methods used in producing modern newsprint, consider using paper printed with the new soy-based inks.

Smoking

On the old homestead, building a smokehouse was third on the list of priorities, after building the house and barn.

That fondness for the rich, smoky flavor hardwood smoke imparts to just about anything that runs, swims, or flies has continued to this very day. Hardwood smoke both gives meat products a rich, full flavor and improves their keeping qualities. The wisps and tendrils of smoke that curled around the buffalo and pronghorn meat hung from the support poles of Native American tepees now flows through electric smokers, which are equally at home on small-town back porches and big-city penthouse terraces.

It does take a bit of our time to properly tend the smoke chamber, and matching hardwoods to create the desired color and flavor of the finished, fully smoked meat products is both art and science. My wife's great-aunt, Nova Warnka, relates an old-time measure to counter insect problems: A bit of sassafras wood added to the smoke generator will act as a natural insect repellent.

Country-cured hams should be unwrapped prior to smoking. Any excess cure or mold growth should be removed with a stiff brush and a rinse in cold water. Smokehouse temperatures for ham should not exceed 90°F — what is commonly called a "cold smoke." The meat should be smoked for about 48 hours or until the desired rich amber color is attained. After smoking, the ham can be rewrapped and hung up for more aging.

Smoking mild-cured bacon. The following five steps are in order:

1. Wash the cured bacon in warm water.
2. Hang the bacon in the smokehouse and allow it to dry for between 48 and 72 hours, with the smokehouse door open.
3. When the bacon is dry, smoke it using your choice of hardwoods to fuel the generator. Use a cool smoke (under 100°F) for 36 to 48 hours.
4. Smoked bacon is perishable, so following the smoking process it should be refrigerated or frozen.
5. Any rind should be removed from the cured bacon prior to freezing.

Sliced bacon can be frozen for two to three months, but for longer storage, wrap and freeze the bacon in chunk-size pieces that can be used quickly after thawing.

Avoid Resinous Woods

Resinous woods such as pine, cedar, or other evergreens must never be used in meat smoking — or even in smokehouse construction, for that matter — because they can impart poor flavor to the meat.

Curing Bacon

As popular and tasty as cured hams are, they quite often have to take a backseat to that center of the great American breakfast, crisp bacon. Bacon is a breakfast staple from Vermont to the high Sierras and, along with mild pork sausage, is one of the products we direct-market from our small swine herd.

Mild-Flavored Bacon Cure

This recipe also comes from the University of Missouri. Use only the freshest of pork bellies, cooled to 42°F, and begin the processing within 24 to 30 hours following slaughter. The cure must be applied within 48 hours of slaughter. The green pork bellies can come from skinned or scalded carcasses. Be sure not to stack warm bellies one atop the other. Trim the bellies carefully.

>   2 **pounds meat curing salt**
>   4 **pounds white or brown sugar**
>   3 **ounces saltpeter**

1. Thoroughly mix the ingredients.
2. Apply the cure by rubbing into the bacon surface ½ ounce of cure for every 1 pound of green belly.
3. Cure the bellies in a well-ventilated room or outbuilding and on a tilted table, so that the moisture produced by and during the curing process will drain away from the bacon.
4. Pile the bacons in a crisscross manner on the table to a height of no more than 4 bacons.
5. Allow the bacon to cure for 7 days. If it should freeze during the curing period, add 1 extra day for each day it might have been frozen. The bacon can then be smoked.

There's Pork in This? — Pork Recipes

The marketing slogan, "Pork — the other white meat," has probably reached just as far around the world as the names "Coke" or "Pepsi," but many folks are still unaware just how versatile pork is on the table. The following few recipes show just what can be done with a bit of pork and a bit of human imagination.

Pork Cake

This recipe comes from family friend, Mrs. Lora Momphard of Silex, Missouri, and the use of sausage, black walnuts, and the wooden toothpick test combine to show its midwestern roots.

½ pound pork sausage
1 cup boiling water
1½ cups brown sugar, packed
1 egg
2½ cups flour
1 teaspoon soda
1 4-ounce package lemon peel
1 4-ounce package orange peel
1 cup black walnuts
1 teaspoon cinnamon
1 teaspoon nutmeg
1 teaspoon cloves
1 teaspoon salt
1 cup white raisins

1. Preheat the oven to 350°F.
2. Mix together sausage, sugar, and boiling water, and let cool.
3. Add the beaten egg to the mixture.
4. Mix together all of the remaining ingredients. Add to the sausage mixture and blend thoroughly.
5. Line a baking pan with heavy brown paper. Add the batter to the pan.
6. Bake for approximately 1 hour and 20 minutes. The cake is done when a wooden toothpick inserted into it comes out clean. Let it sit in the pan for 7 minutes before removing it to a serving dish.

Cracklin' Cookies

From the Lincoln County, Missouri, bicentennial cookbook.

    2 cups brown sugar
    2 cups ground cracklings
    ½ cup whole milk (cold)
    2 large eggs, beaten
    1 teaspoon vanilla
    ½ teaspoon salt
    1 cup ground raisins
3½–4 cups all-purpose flour
    1 teaspoon cinnamon
    2 teaspoons baking powder
    1 teaspoon baking soda

1. Preheat the oven to 350°F.
2. Mix as for any cookies, using the cracklings as shortening.
3. Shape into balls. Flatten out and bake for 10 to 15 minutes.

Cracklin' Corn Bread

From the Lincoln County, Missouri, bicentennial cookbook.

    1 scant cup cornmeal
    Pinch salt
    ½ cup cracklings
    Boiling water

1. Preheat the oven to 400°F.
2. Grease a 9" x 9" baking pan.
3. Stir the ingredients together and add enough boiling water to make for easy spreading.
4. Spread in the prepared pan and bake for about 45 minutes.

Pie Crust

*The roots of this recipe are lost in time, but it is the secret behind more
state-fair winning-pies than you can shake a stick at (as we like to say here).
As a personal note, a favorite boyhood snack of mine was simple strips
of pie crust dough (the scraps from the pie baking) sprinkled with sugar
and baked in the oven at the end of baking day.*

3 cups flour
1 teaspoon salt
1 cup lard
½ cup water

Combine all of the ingredients in a large bowl and mix together thoroughly.

Old-Time Pork Cake

Recipe courtesy of Mrs. Lora Momphard of Silex, Missouri.

1 cup hot coffee
½ pound ground fresh pork fat
1 teaspoon baking soda stirred into 2 cups molasses
1 cup sugar
2 eggs
1 teaspoon allspice
1 teaspoon salt
6 cups flour
1 cup ground raisins

1. Preheat the oven to 325°F.
2. Pour the hot coffee over the pork fat.
3. Combine with all of the remaining ingredients and mix well. Pour the
   mixture into a large bundt pan and bake for 2 hours.

FOUR

FITTING AND SHOWING HOGS

BIGGER THAN A LAMB, but much smaller than a club calf, the hog is one of the most popular of all animals for youth project work. In fact, youth project work is probably one of the most common uses for hogs (after home meat production) on small country places.

Hog production is quite strongly identified with the Midwest and Corn Belt, but many of the major American swine breeds, such as the Chester White and Duroc, were developed in the eastern United States. Hog shows in Oklahoma and Texas often draw entries of 1,000 head or more; in California, too, hogs are a popular project for young people who have very small plots of land from which to work.

A project pig ready for the show ring will weigh between 220 and 260 pounds — and yet it is probably easier to handle in the ring than either a calf or a lamb. Unlike its cartoon image, and as I've said before, a hog is neither a dirty animal nor a glutton. First-year 4-H kids as young as eight regularly show one or two animals in market hog classes in livestock shows from Maryland to Oregon.

Fitting Hogs

In its broadest sense, the term *fitting* is a term used to describe the entire process of readying an animal for exhibition. A hog is "hand-fitted" for an appearance in the show ring. Used in its narrower sense, *fitting* means those final steps taken to get the hog ready just prior to entering the show ring. A washed and brushed hog is a "fitted" hog.

Showing Market Hogs

In youth swine work there are basically two types of projects. One is the market hog project, which is the feeding to market weight of from 1 to 10 head of feeder pigs. The other is the gilt and litter or sow and litter project, in which the young person takes the female from mating to the birth of her litter, and then takes her pigs from birth to market weight or sale as feeder pigs. Both are good learning experiences, though the former is the normal choice for a first-year or beginning youth project.

During my Future Farmers of America (FFA) years, our home-county fair had a daylong breeding stock show for pigs farrowed that spring. Entries would often number several hundred, and some youngsters showed the production from as many as 8 to 10 purebred litters. The market hog class came quite late in the day and generally drew modest numbers. Now breeding stock projects are far fewer in number, and market hog projects have grown more popular. At some county fairs, 300 or more hogs may be entered in market shows and rate-of-gain competitions. At our local fair, the market hog show may take six to seven hours to conclude.

Since they tend to be the more popular today, market hog projects will largely be the focus of this chapter. Information on showing breeding stock, however, appears later in this chapter, and details about raising breeding stock appears in chapters 6 through 8 for anyone interested in litter projects.

Market Hog Projects

A market hog is a barrow or gilt that enters the show ring in the weight range typical of hogs sent to market: 220 to 260 pounds. The typical 4-H or FFA market hog project entails feeding out from one to three pigs with a specific county fair or market hog show in mind. A market hog project has

---

Moving Up to Gilt and Litter Projects

By their third or fourth year in 4-H work, and second or third year in FFA work, I like to see youngsters move up to gilt and litter projects to gain further learning experiences and skills. Granted, the two projects seem to have switched positions in importance over the last few years, and a greater number of young people now participate in the simpler market hog projects.

---

modest space and financial requirements. Good show ring candidates are probably easier to find among hogs than any other major livestock species — which is a testament to the quality bred into today's meat hogs.

Few days are busier, more hectic, or more exciting in a young person's life than show day at the county fair. Months of work and preparation are about to be put to the test, including the young person's selection skill, ability to care for the animal and dedication to that task, and skill as an exhibitor.

Although there are still some spring barrow shows, and some of the national competitions are scheduled in the winter months, most market hog shows fall in the fair season of July through September and most state fairs come from late August through October. The pigs for such shows are born from December through the first four months of the year.

In the East and the Southwest, shows tend to run later in the year, with September and October being the primary exhibition months. The West Coast show season more closely approximates that of the Midwest, with a bit of a later start and an earlier finish. Things wrap up in the Midwest in November with the American Royal at Kansas City, Missouri.

The most common practice in a market hog project is to feed out a hog or hogs for participation in a specific show or event. At the 1994 Lincoln County (Missouri) Fair, 132 4-H and FFA youngsters weighed in nearly 400 shoats of between 35 and 80 pounds. How many of those head made it to the

Stephen August

Few days are busier, more hectic, or more exciting in a young person's life than show day at the county fair.

show ring depended upon such variables as their genetic potential to grow quickly; their physiology and ability to remain sound; how the weather affected their comfort, metabolism, and appetite; how their health and nutrition were managed; and how much time and consideration were given to their care and well-being.

Project Goals

The purpose of market hog youth projects is for the children involved to learn such skills as good animal selection and stock feeding; to gain knowledge of livestock marketing processes; to share in positive group activities with other young people; and to get a good taste of life in the real, very competitive world in which we now all live. I showed hogs and bred purebreds during my own FFA years, have been involved in the purebred swine sector for over 30 years, and led the market hog project for our local 4-H club for several years.

At shows, the animals should enter the ring as good representatives of what market hogs are meant to be. They should be the correct weight, the correct age for their weight, and of good overall type with the correct amount of finish. The market hog is the primary product of the pork industry, and

Hampshire Breed Association

This is a very good Hampshire barrow that is well fitted for the show ring. He has a large frame, good feet and legs, and outstanding muscling.

because of that there is always a great deal of interest in the market hog classes at any livestock show — in fact, they often draw some of the largest crowds of any fair event. Our local market hog show is actually broadcast live over one of the area radio stations.

The shoat in a market hog project has a twofold task: to eat and to grow. It is the young person's responsibility to see that the shoat is kept comfortable, safe, and healthy while fulfilling its tasks.

> ## Have Hogs Show-Ready
>
> The time spent in the show ring is very short when compared to the time spent selecting, growing, and fitting a market hog for its moment on the tanbark. On show day it should thus be market-ready in every way.

The project begins with a rather young animal of just 8 to 10 weeks in age. As the animal grows and matures, its feed and care needs will certainly change. It will be on feed for only 100 to 120 days, making hogs one of the shortest term of all youth livestock projects.

In the end, though, the real goal of any youth livestock project is to produce "blue-ribbon youngsters." As an old-line vocational agriculture teacher once told me, perhaps the best ribbon of all for a beginning youngster to receive is a white one. If this doesn't fill a child with the resolve to work harder and do better next time, then it will serve as the incentive to move on to something else, where youthful talents will be more focused and better applied.

Showing Pigs versus Calves or Lambs

A club calf will cost far more to buy than a project pig (the reserve grand champion market hog at the Lincoln County Fair cost just 76 cents per pound as a feeder pig). A club calf will also be on feed for far longer (and consume much more feed), be more difficult to fit and train for the show ring, and be more difficult to house and transport. Its sheer size can be imposing to many youngsters.

Growing out a club lamb takes a little less time and costs less in feed than raising a pig. Still, the costs to buy are comparable, and in many areas high-quality lambs are not nearly as available as showable pigs. Also, the fitting needs and equipment requirements are greater for lambs than pigs. With a market hog, there is no need for a fitting table, shears, or combs. The basic hog show kit now includes little more than a sprinkler can and a stiff-bristled scrubbing brush.

Housing and Equipment

Housing for hogs being readied for show need be neither expensive nor elaborate. I can recall seeing more than one $3,000 gilt drinking from a $3 rubber pan and snuffling through the dirt of an open lot for stray grains of corn like any other hog.

The facilities for feeding out a small number of hogs outlined in chapter 2 should be more than adequate for holding and readying two or three head for show. Show hogs will be pushed for optimum growth in what can be some of the hottest weather of the year, however, which is an important consideration.

Toward that end, I like to see housing for show pigs placed in an area that is well-shaded by mature trees and that prevailing breezes can flow through. Air flow is critical to cool growing hogs in hot and humid weather; to enhance it, a house with doors on both ends or a back panel that can be raised and lowered is preferable. These can be opened in proper sequence to facilitate air flow through the house and across the hogs.

Hogs being readied for the show ring are pushed for optimum growth in what can be some of the hottest weather of the year. Preferably, their housing will be placed in an area that is well-shaded by mature trees and that prevailing breezes can flow through.

Likewise, roofs that lift or slide back can be positioned to bring more air flow across the hogs. And certainly don't pack the hogs into the housing in warm weather. Allow them something on the order of 12 to 16 square feet each of indoor sleeping space, and that much or more on the outside pen floor. Even better for maintaining soundness and muscle tone in pigs being readied for show is growing them out in drylots with at least 150 square feet of space apiece. To further develop muscle tone, place feeding stations, watering equipment, and sleeping areas well apart from one another.

Once they are penned together, try to avoid moving the pigs to different enclosures, breaking up the original group, or introducing new pigs to that group; any of these can cause stress and injuries from fighting. If the weather should turn very warm, put a sprinkling hose on a timer and position it above the hogs. Set the timer to turn on the hose for five-minute intervals to lightly mist the hogs. Keep this up during the hottest part of the day, generally from 10 A.M. to 4 or 5 P.M.

Feeding for Show

Feed is the fuel that propels a hog's growth. If you are readying a hog for show ring competition, you need high-octane performance.

Toward that end, many producers hold show pigs on a 15 to 16 percent crude protein ration throughout the entire feeding period; I have even seen pigs fed for show on 17 to 18 percent crude protein rations. Even the 3 percent increase in performance that comes from feeding higher-priced, pelleted complete rations is justifiable when readying a hog or hogs for competition, especially if you are going to be participating in a rate-of-gain competition.

Feeding challenges. Going into hot weather, palatability and consumption are two very real problems to be resolved. In very hot weather (temperatures in the mid-80s and higher), the simple act of metabolizing feed already

---

Self-Feeders

A self-feeder should be your feeding tool of choice, because it will keep feedstuffs before the growing hogs constantly, and also maintain the feedstuffs in a cleaner fashion. A small, wall- or gate-mounted, one- or two-hole feeder is probably your best bet, because it facilitates frequent ration changes. You are pushing these growing hogs rather hard for a specific event and so should not stint on either equipment or feedstuffs.

consumed adds to a hog's heat stress and thus decreases further consumption. Rations in hot weather should thus be made as palatable and nutrient dense as possible.

The easiest way to do this is to add fat to the ration. For example, some producers mix in feed-grade fat at the rate of 3 to 5 percent of the total ration. Rations can be top-dressed with an inexpensive liquid cooking oil, or even the discarded cooking oils or greases from a local restaurant. Just be sure to start the hogs on fat slowly and build up its level gradually. A too-sudden shift in ration content can cause gastric upsets, with quite serious consequences for pigs on the fast track to a livestock show.

Pelleting also seems to add something in the way of extra palatability; the heat and pressure applied to the feedstuffs during the pelleting process should improve their digestion and utilization by the animals, as well. Increasing the ration's protein content may also improve growth rates for animals that appear to be bogging down as temperatures increase. Hand-feeding to appetite can also be tried if just one or two pigs are being readied for show. An old trick is to gradually build up the growing ration with milk-rich pig starter until the animal is consuming a pound or so of the creep feed offered to nursing pigs mixed into its ration each day. It's a costly practice, but it's often the best way to achieve much-needed rapid weight gain.

Some producers encounter a problem that is the exact opposite of slow growth: They have a pig growing at a rate that will put it past its prime weight by show time. They, too, will benefit from some hands-on dietary management.

Begin by pulling that self-feeder and going to hand-feeding. Feeding to 90 percent of appetite will help to check growth; some folks pare daily consumption even further. You might also begin by shaving protein levels rather than amounts fed. However, note that in the latter stages of the finishing period you must be wary of either throwing the hog into a complete stall, or slowing its growth curve and causing it to pack on more fat rather than lean gain.

Make any change in the ration a gradual one, and couple any ration trimming with a course of exercise for

Don't Forget the Water!

Do not neglect the most important of feedstuffs, drinking water. In warm weather it is vital that drinking water be kept clean, fresh, and appealing, to encourage maximum consumption. Keep it before the hog in the cooler parts of the day when hogs are fed, which will do much to encourage consumption at proper levels.

the hog in question. Walk the animal for 30 minutes or so in the cool of the day to give it some toning and conditioning.

Selecting Hogs for Show

While a rose may always be a rose, pigs are born quite different from each other. Selecting a pig or pigs for show is not as simple as buying the first thing that walks by with four legs and a corkscrew tail. In fact, there is much more to take into consideration than when you are selecting a hog for market or home processing. Here you are truly trying to skim the cream of the local pig crop.

The best advice I can give on show pig selection is to go out and take in as many nearby hog shows as possible. In particular, try to catch your local hog show judge in action, to learn his or her style and selection criteria. At the shows, note carefully the judge's comments as he or she explains the class placings, and note carefully the animals selected as class and overall winners. A judge doing a good job will talk freely about what he or she is looking for, and the consistency of the judge's opinion should hold up through all of the placings. The 240-pound class winner should look like a slightly older, slightly larger version of the 220-pound class winner.

Sizing Up the Judges

Normally, market hog judges come from one of three groups: purebred swine producers; university staff members or extension service personnel; and representatives of the meatpacking industry such as order buyers. There is

---

Livestock Show History

The livestock show's roots reach all the way back to ancient times, when early nomadic herdsmen would meet to exchange the young of the year and breeding animals to add new blood to their herds and flocks. This tradition continued through medieval market days and beyond. Early in U.S. history there were displays of various livestock breeds and varieties at harvest fairs and farmers' gatherings.

The competitive aspects of livestock showing came later. Over time, the side-by-side comparison in the show ring came into its own as a primary arbiter of desirable livestock type both in this country and abroad.

much to be said for knowing a particular judge's likes or dislikes during the selection process. One of the best examples of this is how the soundness and its effect on mobility are addressed.

Some judges are a little more forgiving of things like uneven toe points or slight defects in leg structure than are others. Don't take such things as givens, however: These are often the same features other judges use to resolve final placings. There's little of economic value in a front leg or a hoof point, but an animal that doesn't move freely and easily won't take its place at the feeder and consume feed and grow as efficiently as it should.

> **Watch, Listen, and Learn**
>
> At the local fair, the place for youngsters to be when they are not actually showing is in the ringside seats, listening and watching as the judge places the other classes in the show.

At many shows judges sign multiple-year contracts; you can thus develop a real feel for two- and even three-year judging trends at such shows. Many states also sponsor early-spring symposiums on livestock judging and evaluation on their ag school campuses; these can be good guides to current trends in popular and winning livestock type. Contact your local extension agent to learn if such events are held in your state.

This young lady has done an excellent job of selecting and fitting her hog for show. The hog is a growthy Spotted barrow that could compete in any show ring.

Tips for Selecting Show Hogs

There is a body of nuts-and-bolts data that can serve you well when you're selecting hogs for a market hog competition. Following is a sample:

- Gilts generally grow a bit more slowly than barrows but nearly always hang a leaner carcass on the rail. Their leaner nature does make it possible to push them harder with a ration higher in crude protein.
- Select the largest and most well-developed gilts in a pen; they demonstrate a higher growth potential than their siblings.
- In some areas market hog competitions are still called barrow shows; castrates are certainly the bedrock on which the modern pork industry is built. To show a barrow, select again from the largest pigs in the pen, but also a trim appearance about the head and along the underline. The presence of at least three nipples ahead of the penile sheath is a good visual indication of an animal's potential to hang a long carcass.
- Crossbred pigs tend to grow faster than purebreds, but to any rule of thumb there are exceptions. Right now, stylish purebreds showing a lot of breed character are a rare enough sight in the show ring to catch a judge's eye and earn a long second look as the classes are placed.
- When selecting crossbred animals, load up on the colored breeding (black and red) as much as possible, to capitalize on the muscling and growth characteristics for which those breeds are noted.
- Black and red hogs appear to show up better under artificial lighting. White or mostly white hogs seem to show better in daylight.
- When possible, weigh in at least one backup pig for each animal that you plan to show.

Desirable Hog Traits

A pig for showing has to be very *typey*. This is a very vague term, to be sure, but what it most specifically means is that the animal demonstrates large amounts of muscling, is very lean and trim in appearance, is of exceptional length, and is otherwise strong in the current trends in swine type.

That latter point, the seemingly ever-varying trends in swine type, is the true kicker in the showing of hogs in any class or category. In this century alone there have been close to 20 major changes in swine type, each one believed to have moved the animals closer to an ideal in meat, reproductive, or growth type. Among the functions of the show ring are to showcase such changes and then to see how well they hold up over time and in head-to-head competition

Prepping for the Show

About a month before the date, preparation should really kick into high gear. **Get the hog used to being handled!** In the cool of the morning or late evening, walk the hog to get it accustomed to being exhibited. Acquaint the animal with both the show stick and the brush. Use light taps on the shoulders with the show stick to keep the hog in motion and turning properly at the corners or ends of the ring. If you tap the pig on the hams or around the loin, it will *unwind,* or go into a sort of clinch then sag or hunch up with its tail uncurled, which is not a pretty picture for the judge.

Get the hog comfortable with being brushed. A brush is the tool for cleaning the haircoat; in the ring brush away bits of dirt, sawdust, or straw from the hog. Older youngsters can often show a market hog with nothing but a brush in the show ring. I don't like to see youngsters over 13 go into a ring using a show stick, hurdle, or cane, because this suggests they haven't spent enough time getting their animals ready. Further, striking, kneeing, or kicking an animal is sharply discounted by nearly all judges.

Begin evaluating the hog's emerging growth and finish curve. Do gains need to be pushed, or does the feed need to be scaled back to produce a trimmer-appearing finish? Twice-a-day walking may be needed to increase trim or tone, and a ration switch as extreme as going down to just a few pounds of shelled corn daily may be in order. To push gains, bring on the pig starter or switch to slop feeding, the mixing of 1 part complete ration to 1–1.5 parts water. This should increase warm weather consumption, add to body fill, and increase finish.

Take an honest look at the hog and seek an outside opinion as to its merit. Ask a project leader, a teacher, extension agent, or local swine producer to help determine show ring strategy and which of the animal's better traits to play up in the ring. A couple of years ago, one of the youngsters in our Silex Flyers 4-H club had a very good Hamp-cross gilt with a bad tendency to bunch up when she stood still. To show well she had to be kept moving in the ring. She was selected reserve grand champion, but had the child not let her get jammed in a corner of the show ring and bunch up in front of the judge, she would have been grand champion. An outsider can help you recognize such problems, along with the means to correct or counter them.

Get cracking on the paperwork. Get in your entry forms and make an appointment with the vet to have blood tests taken so you'll have those health papers in hand before the last-minute rush. You're likely to need written proof that your animal is free from pseudorabies and brucellosis.

with other hogs. It thus becomes imperative that those selecting animals for the show ring be well-read on the subject of swine type, attend a number of shows, and mine their project leaders for as much hard information on swine type and show ring conformation as possible. Hit the swine press hard for information on emerging trends in type and show ring developments.

Leading Show Breeds

At this time, Hampshire-cross pigs stand head and shoulders above anything else in the field of market hog competition. The strongest crosses are the Hampshire x Duroc F_1, the ¾ Hampshire / ¼ Duroc cross, and the Hampshire x Yorkshire F_1. Berkshires and Landraces are also being used in breeding to formulate some snappy genetic combinations for the show ring. A growthy gilt with a lot of length and internal dimension and wrapped in a good black hide is going to be hard to beat in market hog competitions up to about the 260-pound class. Where classes are carried to extremes, up to 300 pounds, the edge seems to tip back to barrows, and also to hogs that carry more white breeding for the trimness they can maintain at such weights.

Going to the Show

At most shows you are expected to have the hog or hogs on the show grounds between 12 and 36 hours ahead of show time. They are generally weighed in as they come off the truck or trailer. That weight is then used to

Market hogs being readied for show should become accustomed to bushing before show time.

determine which class they will be shown in, along with their placement in the rate-of-gain competition if it is to be a part of the show. This is also the time when ultrascan or "Real-Time" readings may be taken on the live animals, to measure things like loineye size and backfat thickness, which are further aids in placing the swine classes.

With fairly heavy hogs, it is a rather widespread practice to withhold both feed and water for a period of 6 to 12 hours prior to weigh-in. This is generally through the night before a morning weigh-in. It puts the hog across the scale with an empty bowel and bladder, and a couple of pounds of body shrink. The animal will regain most of the lost weight as soon as it goes back on feed and water. I neither condemn nor condone this practice; it is very widespread, there are no specific show rules against it at most events, and it seems to have no harmful effect on the animals (feed and water are often withheld from hogs at farrowing for this same length of time).

I do caution against withholding drinking water during very warm weather. And at the opposite pole, keeping lightweight hogs on feed and water right up until they go on the truck may be necessary to make show weight.

Transport

To spare the animal any stress or risk of injury it needs to be a soft, sweet ride all the way to the fair or show. Do the trucking early in the day, while the coolness of the night still lingers. If it is very hot, a 4-inch layer of damp sand or sawdust in the truck or on the trailer floor will help keep the hog more comfortable. A note here: I once exhibited a pair of Duroc gilts that developed an allergic reaction to the oak sawdust bedding used at a local fair. Experiment with any expected bedding changes at home first.

Move the show animal in a very slow and easy fashion. Use no whips or prods, no kicking or kneeing, and move the hog by blocking behind it with a hurdle or short gate. Entice it with feed and let it move along at its own pace; a running hog is a hog in peril of falls and other injuries. And unload it in exactly the same way. That tail is not a handle for lifting and twisting.

Show Box Supplies

As an exhibitor you will also need to take along a show box packed with the essentials for a day or week at the show. This is generally a footlocker- or trunk-type affair made from heavy plywood and painted to look attractive in the back of a pen or along the alleyway in the exhibition hall or show barn. White, blue, red, or green — traditional colors for farm buildings and equipment — is often the color of choice, and many folks further adorn their boxes with decals or stickers of favored breeds and the name of the family farm in large letters. I've seen show boxes large enough that small youngsters can actually sleep atop them.

Into the show box should go:

- ❐ Nested feed and water pans
- ❐ A sack or two of feed
- ❐ A garden hose for watering and washing hogs
- ❐ Brushes for cleaning the animals
- ❐ Dishwashing liquid or other liquid soap to wash hogs (dishwashing liquid will also rid hogs of lice without causing any residue or withdrawal problems if a louse problem should emerge at or close to the show)
- ❐ Folding cot or sleeping bag, if you are required to stay with your animals overnight
- ❐ Show stick
- ❐ A small hurdle or two
- ❐ A pen sign or card to identify yourself

A garden hose for watering and washing hogs, a sack or two of feed, and a folding cot are a few of the items that should go to the show.

Grooming Hogs for the Ring

At one time swine exhibitors applied colored oils to their colored hogs, to bring up the haircoat and to make it glisten in the show ring. White hogs were liberally sprinkled with a white, talc-like powder for the same reasons. Such practices have since been eliminated, to promote livestock shows and exhibitions as showcases for the animals' merits and not the fitters' skills and tricks. I recall once helping a fellow FFA member show his Chester White hogs. On the way to the show ring the white gilt I was handling was liberally splashed with muddy water from a puddle between the barn and ring. The powder on that gilt began to set up in a concrete-like manner. I drove her into a corner of the show ring, knelt down, and began to frantically brush her haircoat. The harder I brushed, the more the powder set up. Just then the show judge gave me a bit of a nudge with the toe of his boot. He pointed to the center of the ring and said, "Young man, we're having the hog show over there." As it turned out, I got into the show and my "dappled" Chester White won a red ribbon.

Soap and water and elbow grease are just about all that are now applied to a hog to prepare it for a show ring appearance. Scrub the pig downward from the centerline of its back. A stiff-bristled laundry brush will do a good job of freeing dirt and manure from a hog's coat. Once you're back in the pen and awaiting the call to the show ring, a sprinkler will hold enough water to keep your brushes damp for last-minute touch-ups.

At most shows the sprinkler can even be carried to ringside, and used to drizzle a little water on the hog's snout as a refreshing measure. Also useful in the ring is a small scrub brush that can be carried in a hip pocket or free hand. Water can be useful in keeping a hog cool — but never apply water to a hog showing any signs of heat stress, because shock and death may be the result.

Attire for Exhibitors

The exhibitor also needs to prepare for the trip into the show ring. A judge I very much respect makes a point of announcing at the start of every show he judges that he will be reading caps and T-shirts. If he doesn't like what they say, it affects his placements.

In the show ring, a shirt or blouse of simple pattern and western cut along with dark jeans or slacks are the best choices for attire. Girls often braid their hair so that it won't be bothersome in the ring. Shoes or boots should be comfortable, support feet and ankles well, and be sturdy enough to provide protection from all those hooves.

Show Time

The long months of selection, preparation, and fitting climax in the show ring, and the action there can be rapid and confusing. You need to go into the show ring with a well-thought-out plan to show your hog to its best advantage:

- Know the pig's strengths and weaknesses and be prepared to put the pig in its best light. Subtly use the show stick to guide the judge's eye to the strengths.
- Do not be the first exhibitor into the show ring, or the last. Try to position yourself to be among the first half of the exhibitors into the ring. Enter the ring with your hog solidly under control and moving in the same pattern as the others there.
- Never come between your hog and the judge.
- The judge seeks views of the hog from head-on, from both sides, and from the rear. To get that head-on view, he or she will be in position to see each hog as it enters the show ring. Some judges even request that the hogs be brought into the ring one at a time rather than in cavalry-charge fashion through a wide-open gate.
- As much as possible, keep the hog's strongest features before the judge. If you have a hog with good hams, set it up for the judge to get a good view of that economically important trait.
- Keep the hog moving and don't get caught in traffic jams in the ring ends or corners. A lot of judges watch very closely as hogs turn, because that is where problems may be most apparent.
- Be prepared to answer questions as to your hog's age, its breeding, and the rations fed to it. These questions may be used to break ties and to determine showmanship placings.
- Follow the judge's directions in the ring closely.

In many shows, the judge will have the top hog picked within five minutes of getting all of the animals in that class into the ring. The good ones show up rather quickly in most market hog classes. The difficult task is sorting through the remainder after the top two or three have been pulled from the group. Deciding where to break between blue and red ribbons and red and white ribbons can also be difficult for a judge. Ranking red-ribbon winners, when required, can be very time-consuming.

Top pigs are generally directed to small pens alongside the show ring, and red- and white-ribbon winners sent back to their respective pens in the barn.

You'll need to keep your hog moving as long as there are animals in the ring.

On the way out of the ring, each hog's paperwork will be turned over to the show secretary, and ribbons will be dispensed.

As long as hogs remain in the ring, though, the show is still on, and exhibitors need to keep their hogs moving in proper fashion. As I've noted, a good judge will give reasons for his or her placements and may even take longer with some hogs lower in the order than with class winners. Some very extreme hogs can present special problems in their placement, and a good judge will address this in his or her remarks.

Show Schedules

Most market hogs show in classes based on their liveweight, with a weight range of no more than 5 to 10 pounds among all of the hogs in a given class. Class first- and second-place winners will be called back later in the show to compete for first and second places in their division. (Generally, a division encompasses the hogs from three to five weight classes.) At most shows there are two to four divisions.

From the division winners are selected the grand champion and reserve grand champion of the show. In divisions and finals, the overall winner is chosen first. After that, the animal that stood second to it in its class or division is brought into the ring to compete with other first-place winners for the second-place position. Staying in the hunt until the last animal is placed can make for a very long day or evening. Our county fair show has often started at 7 P.M. and continued until 1 or 2 in the morning.

With the completion of show ring competition there may be another day or two of show-related activities to work through before a fair or other event is over. There may be a requirement to keep the barns full for later fair visitors, a sale of show winners, or even an auction of all show participants. At many shows all of the animals and gear must be removed, the pens cleaned, and the barn set back as it was before premiums and sale checks are paid to the participants.

Until the show is over and the animals clear of the barn they have to be the young exhibitor's first priority. No matter how much the Ferris wheel and cotton candy may beckon, pens and alleys need to be kept clean, hogs monitored to ensure their safety and comfort, and passersby made welcome. Answer any questions from those visitors; many are potential bidders in the auctions that often follow many hog shows.

Keeping Your Perspective

Sitting astraddle a bale of straw in the back of the show barn there are many lessons to be learned: how to feed a winner, where the best pigs are to be found, how to do a better job of fitting for the show ring, and so much more. And one of these lessons should be kept in mind for as long as you pursue a career of any sort out on the tanbark.

The judge's placing is just one person's opinion. I have seen hogs win a blue ribbon in one competition, then do no better than a red ribbon the following week under a different judge. A properly run show is first and foremost a learning experience: the latest in swine type presented for honest evaluation before the eyes of other swine producers. The ribbons and premiums are but a pleasant by-product — or are supposed to be.

To me, the hallmark of a good hog show is that moment when, on the way out of the ring with a lower-finishing hog, a youngster turns back for a moment to shake hands with the exhibitor of that day's winner. On another day, with another judge, the placements might very well be reversed. The friendships made and the careers launched inside the show ring can last a lifetime, however.

Showing Breeding Stock

Many young people can and do go on to breed and show their own hogs, and even to compete in shows for crossbred and purebred breeding animals. Breeding stock competitions can include county fairs, state fairs, and conference-type competitions at which breeders from across the nation come together to compete.

The basics of breeding stock competitions are the same as for market hog shows, with one exception: Both the handlers and the animals in these competitions tend to be older. Hogs a year or more of age still compete at some state fairs and a few still conduct big-boar contests, which include boars five years old and older, often topping one thousand pounds. "Open classes" at these fairs are open to all comers. Nearly every year at the Missouri State Fair, for example, one breeder or another is honored for 50 years of competition at that grand and classic swine show.

While many of the participants at the state-fair level are veterans of decades of such competition, most state fairs also maintain breeding stock and market hog classes just for 4-H and FFA youngsters. These young people are also generally free to go on and compete in the open classes. By the time the state fairs roll around each year, most of the animals of the year going into the largest breeding stock classes are 8 to 10 months of age.

> ## The Stakes Are Higher
>
> State fairs and national conferences are true arbiters of swine type, and at some a lot of dollars are on the line. A big win can lead to a selling price in the healthy four- or five-figure range for the animal in question. Careers and breeding lines have been launched on just one or two such wins. Such shows are very much merchandising devices for bloodlines and breeders.

Breeding Stock Classes

Breeding stock classes are most often divided by animal age, in roughly 30-day intervals. For example, an early-spring Duroc class might include just boars farrowed between January 15 and February 15 of that year. In open shows there will be breeders who farrow many litters per year and thus can very exactly match animals to each class. Returning to the above example, a boar farrowed on January 15 will be a full month older than the youngest members of its class. Given the rapid growth rate of modern hogs, this can

Spotted Gilt Breed Association

This very feminine Spotted gilt has great body capacity. She can carry a wash-tub full of pigs — demonstrating that the meat breeds can also be strong in the reproductive arena. This gilt was the Hog College Gilt for the Spotted breed at the 1996 National Barrow Shot in Austin, Minnesota, a show that is sometimes called the "World Series of Swine Shows."

provide an animal with a tremendous advantage in the visual appraisal that is still so much a part of breeding stock judging.

Boars about 180 days old are moving into sexual maturity, and one of the most common of the secondary sexual characteristics in boars is aggressiveness. Between boars raised as penmates this is no problem, but boars of this age driven into the tension of a show ring and confronted with other, strange boars can and often do mix it up. A boar fight can be a simple inconvenience or a costly tragedy. More than one young boar has had to be destroyed following a show ring mishap.

The Value of Stock Shows

One of the tests I use to determine the sincerity and dedication of young would-be hog raisers and showmen is to take them to a nearby FFA breeding stock show and sale. If, after seeing one or two teenage boys bounced around by some scrappy 350- to 400-pound boars, they retain their interest in hogs, they may have what it takes to take their own licks in and out of the ring. Actually, I have seen lambs and calves take a harder toll than hogs on youngsters. Still, this is a problem every child will encounter sooner or later.

Not nearly as many seedstock producers take the show ring route now as in the past, but it is still an important venue for swine promotion. The big swine shows are followed and reported upon as much as many sporting events, and the results are as eagerly awaited. Winners at an event such as the World Pork Expo or the National Barrow Show can define the industry for years to come. Within a relatively few months, their sons and daughters can be going into use in herds all across the country.

I'm a true fan of hog shows and especially enjoy seeing breeding stock being shown. There is no better place to practice and refine that all-important stockman's eye. These shows put to a hard test our mental image of a good hog that we all carry around in the backs of our heads. Little does more to raise my estimation of a judge than hearing him agree with my opinions. And little does more to make me a better producer than a judge's comments that cause me to really think about what it takes to make a hog a good one.

In this age of extensive production testing, ever-changing scanning devices, and the use of computers to project estimated breeding values and progeny differences in seedstock, many folks question whether the show ring retains real value in livestock selection any longer. I share a belief with a great many others that, even if it is among the oldest selection and comparison tools available to us, it is still a good way for livestock to be evaluated, and one very much in keeping with the character of the American family farmer.

The show ring provides us with that all-important and objective outside opinion; allows our opinions to be confirmed or disproved by head-to-head competition before the general public; and, in all but rare instances, its animals truly rise or fall on their own merits. In the modern livestock game it has been proven to build character, legends, and herds and flocks that shaped the trade for generations.

Show Ring Controversies

There are many controversial aspects to the show ring. For instance, how meaningful are eyeball comparisons to modern commercial producers? The judging at shows is largely based on eye appeal and does not always factor in things like real carcass data or reproduction performance traits. Are show ring events too competitive and too prize-oriented for young people?

In many instances show ring competition is not, in fact, set up to evaluate such important aspects of swine production as reproductive performance, exact growth rate, feed efficiency, and various carcass traits. And exhibitors with all different levels of experience compete in most local swine classes. It is not uncommon to see an 8-year-old, first-year 4-Her in the ring competing with young people in their late teens with 10 years of show experience. Still, most of those older youngsters began their show ring careers in the same way and all are operating under the watchful eyes of show officials.

It is also possible to "buy" a blue ribbon: Some families have more money to invest in project animals than do others. However, hogs these days are produced in such numbers that pigs of competitive quality tend to be available in many areas for little more than feeder pig prices.

Finally, there are, unfortunately, some show ring cheats. These are exhibitors who attempt to show overaged animals, abuse certain animal health products, such as steroids, or employ the skills of professional fitters and showmen to ready their animals for the show ring. Most abuses can be spotted by show officials, though, and more and more show animals are being forced to submit to health examinations to detect and counter such abuses.

FIVE

THE HOG BUSINESS

IN THE PAGES AHEAD, I will endeavor to discuss the business of hogs and how they earned their reputation as mortgage lifters on farms all across the United States.

At this writing there are a number of mega swine operations dotting the rural landscape. It has even been predicted that agribusiness corporations will gobble up pork production the way they have both egg and broiler production. To present the obvious, however, hogs aren't chickens: They take much longer to bring to a marketable weight, and they're still not being bred with anything near the uniformity found in broilers. Hogs also eat more than chickens, need more space, produce more manure, and require more labor.

Some of the giant swine operations already have begun to falter and fall into bankruptcy; as a group their growth curve has begun to slow, and there are hog industry observers who believe that any investment into such production will be a short-lived one. Indeed, many of these units seem to be starting up with a short-term recovery plan on investment; the thinking is that they will be forced out of business shortly after the turn of the century by humane, labor, and environmental issues.

Twenty years ago there existed a role for specialists who raised nothing but hogs on small to midsize farms. Actually, there were even more specialized roles within that hog-raiser category: Not only were there folks raising hogs, there were feeder pig producers, hog feeders, seed and show stock producers, and even a few experimental breeders. It was probably too specialized an approach to the production of a single agricultural commodity to endure. It isn't wise to rely on a commodity that depends almost entirely on the whims of a wholesale market with a demonstrated desire to reward naught but production in volume.

Small Farms Still Predominate

I'd like to point out the simple fact that the greatest number of sows in the United States are still held in herds of 40 head or less. Even the most ardent supporters of agribusiness concede that at least a good 20 percent of all market hog production will remain in the hands of small farmers for the foreseeable future.

Running a Manageable Operation

All of the reasons to raise hogs that I've cited before remain valid, but these days hogs are best used as part of a truly diversified production plan. When they represent 1, 2, or 3 ventures in a farming mix that may encompass 10 or more such modest-size, noncompetitive activities, they will add real earning power and economic security to the small farm. By "modest-size, noncompetitive ventures," I mean farms with 10 to 20 sows or even less in a venture that nets from $2,000 to $8,000 annually and serves a niche market. Say you have five sows producing F_1 gilts and feeder pigs or butchers for whole-hog sausage; these would be a good fit with a few beef cows and a pasture poultry operation or a pedigreed sheep flock. If we use hogs like our great-grandparents did, their value really shines, whether the family farm is a 1st- or 10th-generation operation and whether it is located in Maine or Hawaii.

States leading in pork production include Iowa, North Carolina, Minnesota, Illinois, Indiana, Nebraska, and Missouri, but hogs can be raised just about anywhere.

Big hog operations certainly seem to get more than their fair share of press attention, and much is made of their so-called efficiency of scale. However, nowhere has that efficiency of scale ever been actually documented. Some believe it may not appear at all until you own sows by the thousands — if then. The big operations have to maintain some semblance of cash flow to satisfy their investors. They are thus locked into producing X amount of hogs per diem regardless of what is happening in either feed or hog markets.

A hog enterprise can fit into a great many small farm mixes without taking labor or resources away from other ventures. Sows can be bred to farrow in any of the 12 months of the year, many hog facilities can be made to do double duty by housing other species (a one-sow farrowing house with a solid floor is also a dandy place to brood baby chicks or waterfowl), and you can fine-tune the management of five sows just as easily as you can that of five hundred.

Getting into the Hog Business

If you want to go into the hog business, the first point to resolve is why and toward what end the hogs are to be owned. A great many small-scale producers make the move to swine production after a few years of feeding out a few pigs for the table needs of family and friends. They have gotten past the myths and misunderstandings about hogs and hog production and found they honestly liked working with the porcine species. They often begin with a gilt or two that had initially been headed to the freezer.

Have Options

A litter of pigs can easily number from 8 to 12 at weaning, and with that many porcine mouths to feed, you must have a clear vision of their intended use and where they are to be sold. Don't take it for granted that you'll always have a wide variety of outlets for your production. Nor can you assume that a niche market now taking 20 head yearly at a fine premium will want or can absorb another 20 head.

We have always lived in Missouri, itself a part of the Corn Belt, and we have always had two of the corn industry's "Big I's" for neighbors: Iowa and Illinois. Still, for a time when I was first home from college, a single large hog feeder — a person buying feeder pigs and growing them to market weight —

would often dominate and dictate the feeder pig market in our three-county area of east-central Missouri. This individual would sometimes buy 150 to 350 pigs weekly and have upward of 2,000 head on feed. Some seasons, this would equal the entire weekly run at one or even two local pig auctions. It would also define the market for several weeks thereafter. Without this individual in the market, feeder pig prices fell by as much as 20 cents per pound in a week.

This points up the danger of relying entirely upon a single outlet and/or a single marketing option. The traditional venture for small-scale swine producers has been feeder pig production. Space and feed requirements are fairly modest, production can be quickly built up to a level that affords at least a monthly paycheck, and sweat equity can replace a great many capital goods in the production of feeder pigs. Alas, the market for feeder pigs may be the most volatile that exists for any agricultural commodity.

> ### U.S. Hog Production
>
> According to figures from the U.S. Department of Agriculture, small hog producers — those with less than 100 head in their herds — account for 62 percent of all hog producers.

Define Your Turf

It is up to you as a small producer to define your own turf and pursue a variety of markets. The stranglehold, for example, of the large feeder I described above was broken when a graded feeder pig auction was established locally, with the support of a number of small producers. Feeders from as far away as central Iowa began to patronize it.

At Willow Valley — far, far from a showplace — we maintain a herd of from three to eight purebred Duroc and Mulefoot sows and gilts. From them, we sell an average of 1.5 to 2 breeding boars per litter, along with a few gilts and some feeder pigs, and we are venturing into the whole-hog sausage/cured-meat business ourselves. Certainly not every purebred pig born is a candidate for the breeding herd, and feeder pig sales are a strong backup market for us at times. Even with small groups of pigs we often get top dollar, because our pigs have a reputation for good performance, and because many buyers are seeking the spare gilt pigs that are left in a group after we pull the keeper boars.

Breeding stock and now sausage are our prime marketing options. We target the boars to other small producers in our area and encourage them to develop pork products for direct marketing in our local East Central Missouri

Farmers' Markets — a group of small farmers working together at several outlets with direct sales to consumers. That is the key, I think, to working with small numbers: Turn their output into premium-quality production, then work to add as much value as possible to it before initiating direct marketing yourself. Feeder pigs are a sort of safety net that can be sold nearly any week of the year at one of a number of nearby consignment auctions. Our emerging pork sausage sales were not a market we originally foresaw, but it's one with every bit as much promise as breeding stock sales.

Finding the Advantage

I am aware of a small producer in Maine — a state not normally associated with pork production — who was initially building a midwestern-style, farrow-to-finish hog operation. He had been building sow numbers and adapting volume production measures until he hit a sharp downtrend in butcher hog prices. A cutback in herd size was mandated, but he was in an area with a very limited potential for the marketing of cull sows.

He turned the old adage, "If you can't sell a commodity you'd better be prepared to eat it," to his advantage. He began turning those sows into whole-hog sausage and marketing it directly to consumers. At last contact, he was trimming his sow herd to 20 females to produce butcher hogs for his steadily developing sausage market.

The Numbers Question

A sort of no-man's-land for pork producers is emerging. Ironically, until a few years ago, most producers were told to aspire to a level of production that incorporated from 75 to 200 sows. But producers with this many sows have hit some very big walls.

Production at these levels can totally dominate a farm, tie up at least one person full time with the hogs, require extensive investments in rather specialized swine-raising facilities, and make the whole farm totally dependent upon the whims of the swine market.

Smaller numbers actually help family farmers to maintain their greatest strength, which is flexibility in the face of adverse circumstances. The 10-sow producer should have the time and space for a number of other small ventures that will give the farm greater economic stability and orderly earnings, fully employ available labor, and add to the environmental integrity of that particular farming unit.

Be a Savvy Shopper

Numerous university studies of swine producers and their farm records have shown that top-ranked producers often pay nearly as much as one-third less for inputs, which would include feed, breeding stock, equipment, and supplies. They use reserves to buy during the off season and seek volume and cash discounts. They do a better job of shopping around and are more current in their knowledge of the marketplace than producers ranked below them. In addition, they tend to show increases in efficiency and output. They are succeeding not because of their numbers of hogs, but simply because they are doing a better job.

I believe this is in part because they begin with an emphasis on quality. (See "Think Quality" on page 112.) They don't stint on inputs to the breeding herd or feedlot group. Good hogs don't eat any more than the other kind and often — very often — they eat less.

What drives the hog industry is the butcher hog market and when it slumps, survival often hinges on the producer's ability to tread the troubled economic waters. The small producer can sell a pig crop as feeders to reduce losses, feed out the pigs if feeder prices are not satisfactory, or even sell out completely and pull the hog houses up around the barn to ride out an especially trying period of distressed prices. Most breeding herds would benefit by trimming away the least-productive third of their members; 60 percent or more can be trimmed away without impairing the solid genetic base needed for a herd to continue.

Conversely, growth need not be driven by outside forces; small producers can build to the level at which they are comfortable without causing competition for available resources with other enterprises on the family farm or business.

The small swine producer has to concede nothing to the big operators or to the segment of the farm media that is so supportive of them. Small producers have access to the same genetics and can still reach most of the same markets. In fact, there are a few niche markets that are uniquely their own. They can tap into the same supplies of feedstuffs, and they can access appropriate technology and data through many of the same outlets that serve the high-volume producers.

Unfortunately, small farmers are too often identified with the wrong kind of production. Several years ago, I was brought up short while strolling through a local sale barn by the sight of a pen of solid black barrows that were crowding 350 pounds in weight. This was also the first whole pen of No. 4 grade butcher hogs that I had ever seen in my life. These are really fat hogs, with carcass lengths of less than 28 inches, backfat toppings of 2 inches, and poor muscling.

Overfinished, small boned, overaged, and just plain wastey, they were the talk of the barn that day. Later in the day, I caught up with one of the auction owners, a man who saw and handled thousands of hogs each year, and asked him about those big, black hogs. He recalled them immediately and told me they had come from a small farmer who sent a couple of bunches each year to that sale.

Think Quality

Such an auction was perhaps the poorest of all possible places to market butcher hogs. It was held in a small sale barn that didn't pay much attention to quality or promotion, with an audience made up of a handful of farmers and traders. Once again, small farmers had taken a direct hit on the quality question because the owner of those fat hogs was perpetuating a negative, stereotypical image.

As the meat markets of this nation become more and more quality driven, the premium for quality will exist for those who can deliver the desired hogs, rather than hogs by the tractor-trailer or pickup load. When hogs don't pass muster on quality, the best thing to do is get them off the farm, take your lumps, and do a better job next time.

With quality in place, your next task is to maintain it at optimum levels of production: the operative term here being *optimum* rather than *maximum*. Toward this end you must shop carefully, invest correctly, and execute all management activities in a truly timely fashion.

Comparison Shop

Small farmers lack clout in the marketplace more often because they think they do than because of the actual conditions there. Comparison shopping for inputs, seasonal buying of inputs such as feed grains, bulk buying, and the like can make for substantial cash savings for even the smallest of operations.

Two elevators in the same town may be as much as 5 to 8 cents apart on a bushel of corn and $1 or more apart on a 50-pound sack of protein supplement. One feed dealer may try to cultivate small accounts by offering services such as inexpensive deliveries, while another may offer an extensive inventory of animal health and specialty products.

Most years, feed grains are lowest in price at harvest. Sixty bushels of corn can be stored in a bin made from the hopper of a discarded 60-bushel self-feeder; 150 bushels will fit into a gravity wagon box set in the bay of a barn; an old chest-type freezer cabinet can hold as much as 1,000 pounds of shelled corn; and a 55-gallon drum will hold 300-plus pounds of the same grain. On-farm storage makes pay-ahead, store-ahead a cost-cutting mechanism for even the smallest of country holdings.

A somewhat opposing school of thought that can make a good statistical case for itself holds that a small producer should buy feed needs weekly, bimonthly, or monthly. In this way, the producer can average out the market highs and lows of the year. It's the way we do it at Willow Valley.

Know Your Market

The successful producer is very attuned and aware. The good ones are well read. At Willow Valley, at least six different swine-oriented publications arrive here monthly, and more are available. Good producers are also plugged into the market outlets that are available to them on a regular basis — not just when they have something to sell — and they are in regular contact with their fellow producers. They study not just the movements in all of the markets that concern pork production, but also what causes those movements to

Traits of the Successful Producer

It's no mystery why some producers are successful. They share several traits in common. They know and honestly like hogs. They also have a clear plan for hog production, emphasize quality over quantity, and have multiple marketing outlets for their hogs. And they develop and enhance these latter three qualities by doing the following:

- ◆ Reading hog industry publications
- ◆ Staying "plugged in" with market outlets
- ◆ Keeping in touch with other producers
- ◆ Knowing what's influencing hog markets
- ◆ Maintaining detailed enterprise records
- ◆ Using those records to make management decisions

occur. They maintain detailed enterprise records and use those records to guide them in making management decisions. They know what it costs to produce a 40-pound feeder pig, a 100-pound butcher hog, or a service-ready young boar.

The good pork producer, then, is a person with both a plan of action and a specific purpose or purposes for his or her operation. It's a person who farms ideas first with a pencil and paper; if the ideas pan out there, then — and only then — does that producer invest sweat equity and precious capital into them.

Feeder Pig Production

Feeder pig production is considered to be the livestock venture with the fastest potential return on investment. Within 170 days of the first females being bred, you can have something to sell: lightweight feeder pigs.

A feeder pig or shoat is a young hog sold to another farmer, who will feed it to a desirable market weight. Earlier in this century feeder pigs were often called "cornfield hogs," a term in common use even when I was a boy. Back then they were young hogs large enough to be turned out to glean the cornfields following the harvest, or simply to harvest the whole crop of corn themselves. This latter process is known as "hogging down" corn.

Friend Ron Macher, publisher of *Small Farm Today* magazine, advocates and follows the hogging down practice even today. He will fence off an acre parcel of his open-pollinated field corn and let 8 to 10 head of growing hogs harvest their own feed grain. The open-pollinated corn is generally several percentage points higher in crude protein than today's hybrid varieties, and for Ron this is a highly cost-effective means of both corn harvest and swine production. He turns the largely self-reared hogs into high-value, whole-hog sausage. He also benefits from the manure they apply to his crop ground; the corn generally leaves the farm as seed corn that sells for $1 per ear at trade shows.

The modern feeder pig tends to average lighter in weight than its cornfield counterparts of yesteryear. You will still see a few 100-pounders pass through the sale barns, but most feeder pigs now fall into the 35- to 70-pound weight range. They are typically between 8 and 12 weeks of age.

Ideally, feeder pigs move directly from the farm on which they are farrowed and weaned to the farm of the feeder, who will take them all the way to market weight. More often, however, they pass through an auction and/or an order buyer's hands. The latter groups and delivers feeder pigs to farmer and feeders on a commission basis.

Goals for Feeder Pig Production

Feeders seek pigs that are ready to go on full feed; that are able to use self-feeding and watering equipment; and that will grow rapidly and efficiently. Toward that end, the farmer/farrower should produce pigs that are:

1. Bred to be fast growing and of good meat type. Most feeders now prefer a crossbred pig of at least 50 percent colored breeding. Using a boar that has available performance data on traits such as loineye area, backfat thickness, and days of age to 230 pounds will help to make his pigs' performance more predictable and is a good selling point for them. A three-breed rotation, such as Hampshire/Yorkshire/Duroc, Hampshire/Chester White/Duroc, or Berkshire/Yorkshire/Duroc, will enable the farrower to produce typey feeder pigs while also being able to retain his or her own female herd replacements.

2. In droves of 35 to 60 head. Feeder pigs in larger droves can bring up to several dollars more per head than pigs that come in small numbers or odd lots, no matter how high their quality. Most pens for finishing hogs are centered on a self-feeder with from 8 to 12 feeder holes or lids. The rule of thumb is that each of those feeder holes should support from three to five head of growing/finishing hogs. On the lower end, this is doable with five sows bred to farrow together. A half-dozen sows should ensure goodly numbers and form a neat-size production unit that will fit into a single sleeping shed, even with the herd boar included.

3. Similar in type and color and within a couple of pounds of each other in size. Eye appeal does much to sell feeder stock, and one of the things most sought after is an even appearance among the pigs in a drove.

4. In good health. Bright haircoats are an indication of freedom from parasites, for example. Clear, bright eyes and an alert manner are also good indications of overall well-being. Any barrows in the group should have completely healed from the castration procedure.

5. Well over the stress of weaning and prepared to fend for themselves on the feeding floor. The temptation is to sell the pigs right off the sow while they are still milk-slick and have their baby bloom, but pigs that walk the fences of their new pens squealing for Mama instead of heading right to the feeding and watering equipment will not earn the producer many repeat buyers.

What a feeder seeks in the way of desirable feeder pigs includes good health and large enough numbers that all of the pigs in a finishing pen will be from a single source. A single source helps reduce problems with disease and ensure that the pigs will be the same size for finishing. Pig herds also have a "pecking" order, and if pigs come from the same source, their place in the herd is established; pigs from multiple sources are likely to fight and bite while they establish their place in the new herd.

A set of sows will produce 2 to 2.5 litters of pigs per year. Thus, with just five one-sow farrowing houses, for example, you could produce at least 40 feeder pigs to sell every other month from a herd of just 15 sows and one boar. The highest-selling feeder pigs are generally farrowed in the cold-weather months of December, January, and February. Late-spring- and early-fall-farrowed pigs generally sell in what are seasonally depressed markets.

Still, there is something to be said for following a seasonal approach to farrowing. There are no supplemental heat costs, it can often be carried out in simple housing on pasture, feed costs may be somewhat less, and housing needs are both simpler and less costly.

Feeder pig production is often called the "foot-in-the-door swine enterprise." Startup costs are modest: You can begin with a scant handful of gilts that may even have been bought as feeder pigs themselves.

> **Reputation Counts**
>
> The one best selling tool for feeder pigs is the seller's reputation for dealing fairly and selling pigs that get the job done and make money for their new owners.

Farrow-to-Finish Production

The most common swine production option pursued in the United States is the complete route, commonly termed farrow-to-finish production. It is more costly than feeder pig production but has been demonstrated to be the most consistently profitable type of swine production.

The farrow-to-finish operator owns the pigs from birth to slaughter. This process requires the producer to invest in an additional set of growing/finishing facilities, the hogs will be on the farm for an additional three to four months, and the growing hogs will consume substantial amounts of feed grains.

The small-farm farrow-to-finish operator normally works with just one or two sets of 4 to 10 sows each. With the sow groups bred to farrow at 60-day intervals, they can easily be worked through a single set of farrowing huts, and farrowing in really severe weather can often be avoided entirely.

Some might choose to feed out one set of pigs and then market the next as feeders, or even split-market each farrowing. A pen of growing/finishing hogs can really tax a small farm's resources, more so than other farming ventures. To reach a market weight of 230 pounds, each will consume $50 or more in feedstuffs over and above what it received as a young pig, and will require an investment in labor of at least 1½ hours to market, along with a rather specialized set of facilities.

A number of small finishing units can be bought or built on runners or old mobile home frames to hold 25 to 65 head of finishing hogs. They will represent an investment of $2,000 to $4,000, not including feeding and watering equipment. These units save a great deal of space on the small farm and will qualify for investment credits under current tax laws.

weanling or lightweight feeder pig

a shoat at about the midpoint of the finishing period

a market-weight hog

Raising pigs from farrow to finish means taking them from birth to slaughter.

Costs of Hog Production

Since costs don't vary much from year to year, this chart should give you a pretty good idea of how your costs will differ with different types of hog production. (Adapted from Harvest Publications.)

Inputs	Producing Feeder Pigs (sows bred mid-Apr.; pigs to be sold Dec.)	Farrow to Finish (sows bred mid-Apr.; pigs to be sold Feb.)	Finishing Feeder Pigs (pigs bought mid-Apr.; marketed Sept.)
Interest rate	9.75%	9.75%	9.75%
Labor @ $7.50/hr.	$67.50/litter	$90/litter	$7.50/hd.
Vet. med. facilities operating cost	$40/litter	$85/litter	$4/hd.
Marketing cost, including hauling	$12/litter	$15/litter	$2/hd.
Depreciation, interest, tax, and insurance on buildings and equipment	$100/litter	$150/litter	$6/hd.
Cost of 40 lb. pig	—	—	$50/hd.
Feeding period (days)	—	—	145
Number of pigs sold per litter	7.27	6.97	—
Market weight (lbs.)	40	230	230
Feed: Corn (lbs.)	1341/litter	5314/litter	570/hd.
Price/bu.	$2.75	$2.75	$2.75
40% supplement (lbs.)	225/litter	1131/litter	130/hd.
Price/ton	$361	$361	$361
18% pig starter (lbs.)	78/litter	78/litter	—
Price/cwt.	$20.25	$20.25	$20.25
16% pig grower (lbs.)	273/litter	273/litter	—
Price/cwt.	$12.25	$12.25	$12.25
Price you'll need to:	(per head)	(cwt.)	(cwt.)
A. Cover feed and variable costs	$30.00	$40.50	$41.50
B. Cover total costs (breakeven)	$45.75	$50.75	$54.25

Only major inputs used in above calculations are listed. Final figures are rounded off to the nearest $.25. Each 10 cent per bu. change in corn price from $2.75 moves total breakeven cost 50 cents in the same direction. Interest rate for finishing feeders based on special rate for purchasing animals.

Note: You might extrapolate from this chart the value of Boar A, which maintains current average pig days to market weight; Boar B, which shaves off two days; and Boar C, which shaves off three days. The savings on a 10-sow herd producing 170 market hogs yearly would be the margin to use for buying a better boar. The same comparison would be possible with boars that might shave .15 and .25 pound off feed efficiency.

Finishing Hogs

Finishing used to be called "fattening hogs" before we hog producers got image conscious, and with the hogs of 50 or 60 years ago it was a quite literal description of the process. Buying feeder pigs and taking them to a good market weight is a costly venture with a substantial amount of risk. It is commonly pursued by high-volume grain producers who are feeding hogs to have one more grain-marketing option.

Finishing feeder pigs is a far less common venture among small farmers, but many pursue it on a small scale as a part of a niche marketing venture. They may feed out small numbers of hogs for sale directly as processed-pork products such as whole-hog sausage, custom-fed freezer pork, or hogs for roasting, or for sale to other specialty outlets.

As a small-scale finisher, you can often find bargain pigs by buying small or irregular groups of feeder pigs that pass through auctions in the wake of the bigger, more desirable droves. If you're equipped with adequate facilities for housing them, you can have some real bargains by buying lightweight pigs in very cold or inclement weather. One veterinarian in our area paid for a good part of his school expenses by buying ruptured pigs — that is, pigs with a hernia of the umbilicus or scrotal variety — along with pigs with downed ears, boar pigs, slop pigs, and the like. He provided health care at cost and then fed the low-cost pigs to good slaughter weight.

Sources of Growing/Finishing Pigs

The most common method of bringing in pigs for the grower/finisher phase of production is on-site farrowing, according to a 1995 government study. Only 10 percent of swine operations used off-site farrowing and nursery units as a source of pigs, while nearly 14 percent obtained pigs from feeder pig producers and just 5.9 percent purchased pigs from an auction, sale barn, or livestock market.

Breeding Stock Production

Producing breeding stock, which you'll learn a lot more about in the chapters ahead, is also an option for the small-scale producer. Contrary to popular belief, you do not need sows by the hundreds to produce boars — which are still largely sold just one or two head at a time. A few years back, two of the three division winners in the Hampshire show at the Missouri State Fair came from a single Hampshire sow herd with just five sows to its credit.

If they are of good type and bred to an equally good boar, a group of five females can easily produce 8 to 10 breeding-quality boars to sell twice a year, along with a goodly number of quality siblings that can be sold in other ways. Those same five females could also produce two 10- to 15-member groups of F_1, crossbred replacement gilts annually. F_1 means the first-generation off-spring of a cross between animals from two separate pure breeds; such offspring have hybrid vigor.

Over the years we have owned a number of sows that regularly produced two to three keeper boars per litter. Even a quick reading of breeder sale catalogs will show sows producing as many as six marketable boars in a single litter. We once owned a Hampshire sow that produced 35 marketable boars during the five years she was in our herd.

Purebreds

In the swine sector, both purebred and crossbred breeding animals have substantial value. With purebreds, the primary market is for boars to be sold to commercial producers, who use them in crosses to produce feeder pigs or butcher hogs. They are typically sold as 8- to 10-month-old boars ready for light service.

The farm press is filled with stories of pedigreed boars selling for four and five figures at events like the World Pork Expo or summer conferences, but the greatest number of boars now change hands in the $200- to $600-per-head price range. Still, that is at least twice the price of a butcher hog, which costs about the same to grow out.

A few years back, when small, commercial sow herds were kept in a more varied state, there was a bit more demand for purebred gilts, and some people held that the real profits in a purebred operation stemmed from gilt sales. It may well be that with some pure breeds now, the gilts have better profit potential if sold at feeder pig weights as youth project animals. Gilts from the white breeds are used by some commercial producers in what are called parent and grandparent matings. They are used to produce in-house F_1 and F_2 herd replacements for the sow herd.

Crossbreds

Crossbreds — F_1s to be precise — have value as both breeding boars and replacement gilts. The boars carry a blend of breed strengths as well as packing that always desirable hybrid vigor. They are perceived to be more hardy and vigorous in the breeding pen than their purebred counterparts. The F_1 gilts can be bred to a boar of a third breed not present in their cross and produce crossbred pigs with good hybrid vigor and reasonably predictable genetics, and their own gilts can be brought back into the breeding herd without disrupting breeding patterns and heterosis levels.

Commercial gilts normally change hands at 250 to 300 pounds and at six to eight months of age (when they are ready to breed or close to it). They typically sell for a $25- to $100-per-head premium over and above what they would bring as No. 1 butcher hogs. Their price base is generally pulled from one of the stronger midwestern markets, such as Omaha or Sioux City. The boars generally sell in the lower range of purebred boar prices, with most going for between $250 and $400 per head. For some producers, they are the Chevrolet alternative to the purebred Cadillac. Without good purebreds, however, they would not be possible, so there is a place for a good purebred boar in even the smallest commercial swine herd.

At Willow Valley we have sold a number of Duroc x Hampshire boars over the years and found them to be nearly as popular as our purebred Duroc boars. Still, they were at best a compromise between the two breeds, and not as strong in the traits most commonly identified with their parent breeds as purebred representatives of those breeds would have been.

Popular Crossbreds

A new generation of crossbreds that are finding favor are the ¾/¼ crosses, such as the now very popular ¾ Hampshire/¼ Duroc boars that result from the cross of an F_1 Hampshire x Duroc female to a purebred Hampshire male. They have a color pattern similar to off-belt, purebred Hampshires and tend to have a very high degree of performance predictability.

On the gilt side, the most popular crosses are the Hampshire x Yorkshire, Hampshire x Landrace, Hampshire x Chester White, Duroc x Yorkshire, and Hampshire x Duroc females. Bred back to a colored male of a third breed, they should produce market hogs that grow well and yield good quality.

With such crosses, white breeding can be introduced into a herd or reinforced without tipping the scales too greatly away from the meat-type genetics that predominate with the use of the colored breeds.

Breeding Stock Producers

Breeding stock production takes place at various levels. It is rather like a three-tiered pyramid. At the top is a rather small group of producers who shape most of the trends in breeding stock types, and market primarily to other purebred producers.

The second layer is sometimes termed "multiplier breeders." They may work with multiple breeds, purebreds and crosses, or all crosses. They are volume producers seeking to sell to the larger commercial producers. They can supply 100 gilts for repopulation of an entire herd, or 20 boars that can work together in a group. Some of these producers may still show hogs, but their primary marketing tool is performance data, which they use to formulate estimated breeding and performance values for the breeding animals they sell. These estimated values allegedly make it possible to project to what level a trait can be improved through a specific animal. A given male might thus be able to shave one day off the existing herd's average days to market.

The larger the tested populations, and the more testing that is done, the more precise these estimated projections will be. Much hinges on the exactness of the testing procedures, the honesty of the test data, and the size of the test population (the larger the size, the better the data). To me, this procedure seems to lack adequate safeguards, be costly to carry out, and — like much of the effort that went into amassing testing data in the past — often do little more than discourage or drive out small and midsize family farm breeders.

A third group pursues purebred production and use on a more traditional level. These producers may raise just a single breed of swine, and they quite often operate with 50 sows or less. They are the ones helping to maintain a number of the minor and endangered swine breeds, along with the other traditions of the family farm. Their primary market lies within 50 to 75 miles of the front gate, they sell a lot of $200 to $300 animals, they produce purebreds but without recorded pedigrees, and they sell hogs in small numbers to other small and midsize producers.

Looking Ahead

Before we proceed to the chapters on acquiring and managing breeding stock, let's take a look at the future. The breeding stock industry now relies almost entirely on the Duroc, Yorkshire, and Hampshire breeds; and to a lesser extent on the Landrace. Many believe that the remaining pure breeds and their backers will have to follow the lead of the sheep industry, where a large number of pure breeds have been taken up by true family farmers who maintain small breeding groups as part of truly diversified farming programs. They support a number of breed groups and shows, are very supportive of each other, promote purebreds even as market animals, and sell their breeding animals to other small producers.

According to this view there will be fewer very high-selling hogs in the future, but as the breeds are cultivated for their individual strengths they will reaffirm their rightful roles as building blocks within the pork industry. The trend toward gourmet pork, such as what is now growing up around purebred and high-percentage-bred Berkshire market animals, may continue. Such animals' meat has a texture and cooking qualities highly esteemed in the Orient, and they are now beginning to be recognized here.

Check the roots of most of the great and lasting purebred herds and flocks in this country and abroad and you will find that they extend back to just one or two foundation females. This points up that while it is a challenge to launch a lasting purebred operation, it is also doable on a modest, truly human scale.

We bought our first purebred gilt for $117.50 at an auction conducted by local FFA youngsters. Our first purebred boar was farrowed in the stripped-out hulk of a 1953 Plymouth. When we bought him we were driving a 1952 Plymouth. On the way home Dad began laughing — it was the first time he had ever known of a hog having to come down in the world.

As in nearly every endeavor in life, agricultural or otherwise, start small, grow as you learn, emphasize quality, and treat everyone in the exact same way that you like to be treated. An unwritten but ironclad rule in the production of swine breeding stock is, "Sell only the kind you like to buy."

Six

PREPARING FOR YOUR BREEDING HERD

I'VE ALWAYS BELIEVED IN HAVING A PLACE TO PUT A HOG before you go about acquiring one. It makes life so much simpler, and it keeps your boots and jeans oh so much cleaner.

To house a sow herd and breeding boar, you will need two sets of facilities. One will be used to contain the sows during breeding and gestation, the other for farrowing and pig rearing. Your housing and fencing choices will hinge on the approach you take to swine production.

On the small farm, sows are generally maintained in drylots, on pasture, or in a combination of the two. At Willow Valley we use drylots to contain our Duroc and Mulefoot sows. The ground is hilly, and in a normal year we receive 35 inches of rain or a bit more; the sloping lots drain quickly and reduce problems with mud. In flat lots, huge mud holes can form during rainy seasons, especially around feeding and watering equipment.

The Drylot

In a drylot we try to provide at least 250 square feet of lot space per sow. On flatter terrain or where rain amounts are greater, sows may need two or three times that amount of lot space. The drylot system lends itself especially well to hilly parcels, wooded lots, and those little, odd areas not convenient or suitable for cropping or grazing. Along wooded fencerows, field margins, and rolling parcels just going to scrub growth, hogs can neatly fit in, be sheltered from sun by the shade that trees provide, and yield a good economic return on a parcel that might otherwise produce no income.

One of our drylots was in continuous use for nearly 10 years and still sheltered sows in a safe and healthful environment. However, many producers prefer to rotate drylots every few years, plowing them up and sowing them to grass and/or legumes for a time. Leaving them idle for 12 months should disrupt the life cycles of a great many harmful parasites and disease organisms.

Environmental Safeguard

At the foot or bottom end of each drylot, we maintain a strip of sod 10 to 20 feet wide. This naturally filters the runoff from the lots and thus serves as an environmental safeguard. A hard rain clears the lots to what is essentially a new surface, and the sod maintains the lots in a natural way, with minimal runoff getting past the strip and little or nothing in the way of erosion in the lot.

Following the very wet years of 1993, 1994, and 1995 we idled one sow lot; within 60 days it had naturally sprouted a number of plant varieties, something quite exceptional, as many veteran swine producers will concede. Wastes both in drylots and on pasture perk through the ground and grass filters naturally. On the smaller farms I know of, most manure is handled in spent bedding, and goes onto crop ground and pastures in early spring, ahead of tillage.

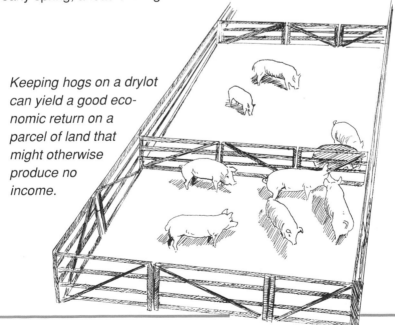

Keeping hogs on a drylot can yield a good economic return on a parcel of land that might otherwise produce no income.

The Pasture

Few sights are prettier than a set of hogs, slick and shiny, on a rich, green pasture. It is a picture of health and wholesomeness. Hogs can fare well even on total legume pastures — something few ruminants can do.

Hogs are not the most efficient users of pasturage, however. They are omnivores and have a single gut. They do not totally utilize browse — twigs, leaves, and shoots — and they need richer sources of energy. One-quarter acre of pasture will adequately carry up to four sows with their litters, but it has to be considered little more than a dessert option for those sows. Nursing sows on pasture must be given a full-fed lactation ration. Gestating sows on good pasturage might have their rations trimmed by ½ to ¼ pound of protein supplement and 1 to 2 pounds of corn daily. You need to closely monitor sows for fleshing and tone at all times, whether they are on pasture or not.

During the last one-third of the pregnancy, sows on pasture should be fed virtually the same as if they were in a drylot. One acre of good legume pasture will supply something on the order of one-quarter of the nutritional needs of 10 head of growing/finishing hogs. Finishing hogs on pasture is probably not the most efficient use for today's higher-priced grazing and farming lands.

Donald Bixby/American Livestock Breeds Conservancy

Finishing hogs on pasture may not be the most efficient use of land in areas where grazing and farming acreage is high priced, but it can still be profitable — and it sure is a pretty sight.

The Pasture Crop

The pasture crop that would absolutely stand up to the wear and tear of grazing hogs would have a great deal in common with Astro-turf. Alas, such a species does not exist. Still, hardiness and durability are important factors to take into consideration when selecting plant varieties for swine pastures. Some of the hardier alfalfa varieties, clovers, and various grasses can be used to pasture swine. Multiple varieties create a more enduring pasture; mixes also are often less costly to establish and maintain than pure stands, especially pure stands of some of the legumes.

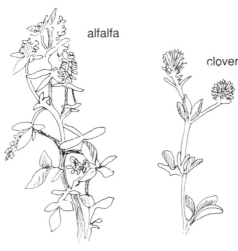

alfalfa

clover

Hardier species of alfalfa as well as clover can be mixed in with various grasses when raising hogs on pasture.

The Central Hub

A relatively newer approach to rearing hogs on pasture is to maintain them in a series of pens arranged in a wagon wheel pattern off a central hub. Access points to the pens are at the hub, as is the primary drinking water source. Also maintained at the central hub are feed storage facilities, handling and loading facilities, and cold-weather farrowing and nursery units.

This was the approach used a couple of decades ago in the Missouri Ozarks, when this area was the feeder pig capital of the United States. Breeding, much of farrowing, and grow-out were accomplished in the rolling, wooded lots. Herding dogs were very valuable in these operations.

The wagon wheel system greatly simplifies caring for hogs on range, reduces the need for costly service and access roads and lanes, and makes it possible to extend a single water source into a number of lots and pastures.

Around buildings and in areas where hogs are to be worked or sorted, I prefer panels, gates, or woven wire supported by posts set on 8-foot centers. The fencing material should be placed inside the fence posts, so that there is no risk of the animals pushing it off the posts by rubbing against it.

When the central hub approach is used, hogs are maintained in a series of pens arranged in a wagon wheel pattern. Access points to the pens are at the hub, as are the primary drinking water source, feed storage facilities, handling and loading facilities, and cold-weather farrowing and nursery units.

Protecting Pastures

Hogs on pasture should be kept rung to minimize rooting damage to pasture crops and the soil surface. Such activity can kill valuable plant species and set the stage for erosion problems. Ringing is the clipping of soft metal rings to the top rim of the hog's nose or across the end of the nose with what is called a "humane ring." The ring causes mild pain to the animal when it tries to root, and thus discourages rooting activity. The rings are inexpensive, simple to attach with a plierlike tool, and cause no lasting pain or disfigurement. On the other hand, they are also easily lost (especially those in the tip of the nose), application requires the hog to be restrained, over time they may have to be reapplied, and there is the pain factor, which is troubling to some people. Still, there is very little in the way of alternatives to ringing other than putting the hogs up on concrete.

Hoof action can also take a toll on soil surfaces. Feeding and watering equipment should be placed on runnered platforms or concrete pads. The runnered platforms, made from durable and inexpensive native hardwood such as oak, can be towed about to prevent problems with wallowing and mud buildup from developing. Placing the platforms or pads immediately adjacent to fence lines will make over-the-fence filling possible and greatly simplify feeding and watering chores.

Hogs on pasture should have a humane ring on their nose to keep down rooting damage to pasture crops and the soil surface.

The Profitability of Pasture versus Drylot

To many, the idea of raising hogs on pasture seems as dated as high-button shoes or buggy whips. Yet in many parts of the Midwest where crop yields are legendary and land prices regularly top $2,000 per acre, pasture hog rearing remains not just popular but also profitable.

In many counties of western Illinois, southeastern Iowa, and elsewhere, sows on pasture continue to thrive in the face of an industry seemingly bent on building ever-bigger, ever-more-costly buildings for the confinement rearing of swine. The pastures are a part of a successful and longstanding crop rotation system; the hogs return manure, urine, and spent bedding to the land; and per-sow investments on these farms are measured in tens and twenties rather than hundreds — often as many as 30 hundred dollar bills per sow in a state-of-the-art confinement facility.

Rotating pastures after 12 to 18 months of use breaks up parasite and disease organism life cycles. Tillage will then eliminate any rooting or wallowing damage. Such farms thus have a margin of sustainability few others can ever hope to equal.

Hogs, however, take a toll on growing plants due to the nature of their hooves and foraging instincts. A drylot will simply support greater numbers of hogs per acre. To maintain a rather delicate pasture plant mix, you have to match the number of hogs with the available pasture. Hogs can also be held for longer in drylots before lot rotation is needed, and are generally more accessible in a drylot. By using drylots we have maintained five to eight healthful and content sows on our 2.86-acre farm — something we could not have done in a full pasture situation.

Fencing Choices

Whether it's on drylots or pasture, the area you enclose for your hogs will be far greater than the few square feet needed for a growing animal or two on a slatted floor. Hogs must be contained in a manner that holds them safely but does not bust your budget.

I am from a part of the country and a tradition that holds that the very best fences are bull strong, baby chick tight, and horse high. Unfortunately, such fences are now dearer than diamonds to erect. Further, hogs normally breach their enclosures by going under them or, as Dad used to say, finding a little hole and worrying it into a big one. You are indeed trying to contain an animal with a bulldozer blade for a nose.

Hogs can be successfully contained with woven wire, a combination of woven and barbed wire, gate-type panels, or electric fencing.

Woven Wire

Woven wire in the 26- to 34-inch-high range is what is commonly called hog wire. It is high enough to discourage most hogs from jumping over it, is easily handled (the 26-inch rolls can be wound and unwound within the space between two rows of corn), is quite durable, and can be supported by wooden or steel fence posts. It is by far the best choice for perimeter and other permanent fences.

It is also the most costly among swine fencing options — as well as the most complex and time-consuming to erect. Setting woven-wire fences is one of the big jobs of our year, reserved for the summer months between planting and harvest.

Woven wire for swine fencing should be tautly stretched. Good corner posts set deeply (30 to 36 inches, to get below the frost line), braced in every direction from which the wire will be drawn, and selected for long life in the fence line will ensure fencing that is of the greatest durability, safer to erect, and easier to maintain.

Hogs can be fenced in with woven wire, a combination of woven and barbed wire, gate-type panels, or electric fencing. But woven wire, pictured here, is by far the best choice for perimeter and other permanent fences.

Barbed Wire

Woven wire as a fencing medium can be enhanced and strengthened with the addition of four-point barbed wire. Positioning the woven wire 4 inches above the ground with a strand of barbed wire beneath it will create a longer-lasting fence line, and one that will discourage hogs from rooting beneath it. Two more strands of barbed wire 4 and 8 inches above the woven wire will discourage jumping and make the fencing usable with other species of livestock as well.

Hog Panels

Hog panels in some ways resemble woven wire, but are made of much heavier-gauge materials and measure 34 inches high by 16 feet long. Fastened to wooden or steel posts at 8-foot intervals, they make a durable, although rather expensive, fencing option. They can be attached to the posts with long-shanked staples, clips, or short lengths of tie wire. The panels can be taken down for reuse and can be handled by just one person.

These panels are also useful for forming simple gates and spanning water gaps. Wired up in such a fashion they make what Dad used to call "pack gates." To open them, you have to lift up one end and pack it around until the desired space in the fence line is opened.

Corner Posts

When I was a youngster, we used for corners a lot of cedar posts that were 8 to 10 feet long and at least 8 inches across at the small end. Dad liked to have his corner posts set in place for a year before attaching fencing to them, to be sure they would hold his fence wire fiddle-string tight. Posts left in for a while settle in the earth and are sturdier; if you fence to posts too soon, they'll shift around, and you'll have more fence maintenance work to do.

A Missouri fencing trademark was and is 8-foot-long crossties set 3 feet into the ground in a concrete footing for corner posts. I believe that some of those old creosote-stained monoliths will still be standing when the Gateway Arch of St. Louis is but a memory. Wooden corner posts should be at least 8 inches in diameter, and be pressure treated or cut from a long-lived wood such as red cedar or locust.

Double-bracing corner posts with treated poles or timbers will further strengthen their holding power. There is also now a system that makes it possible to double-brace 7-foot-long steel posts with other steel posts and use them for solidly anchored fence corners.

Line Posts

Line posts need to be set at 10- to 15-foot intervals, with the longest intervals used in the long, straight stretches. With frequent or sharp dips along the line, fence posts need to be placed closer together, to keep the fence pulled down to the proper height.

Wooden line posts should be set at least 18 to 24 inches deep; they should be at least 4 inches in diameter. In among the wooden posts should be a few steel posts, to ground the fence line in the event of a lightning strike.

Five- to 6-foot-long steel posts can also be used as line posts. They should be set at the same intervals in the fence line as the wooden posts.

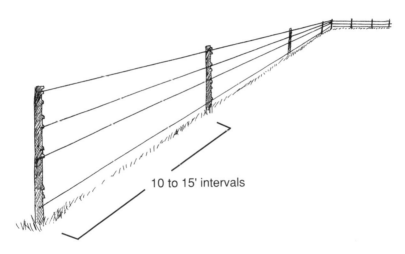

10 to 15' intervals

Line posts should be set at 10- to 15-foot intervals, with the longest intervals used in long, straight stretches.

Electric Fencing

The fourth option for containing hogs is electric fencing — "hot wire." It is the least-costly fencing option, it goes up quickly and easily, and you can hang a couple of miles of it on the barn wall. It can safely contain hogs of all ages and sizes, but is not a good choice for perimeter fencing because it is the most easily breached of all of the fencing choices.

Chargers. The fence receives its charge from a boxlike transformer commonly called a charger. There are chargers powered by electrical current, batteries, and solar power. Electric fence chargers are one of the few things left in this world that operate for a few pennies a day.

Chargers need to be protected from the elements, which can be done either by installing them within an outbuilding or by covering them with some kind of weatherproof box or cover. I've seen chargers wired to a fence post, plugged into a $3 imported extension cord, and protected by nothing more than the bottom of a $2 Styrofoam cooler give years of service, but it sure ain't professional.

With the charger placed in a building, the charging wire can be passed through a hole bored in a wall and insulated with a short segment of rubber hose or tubing; run underground and insulated in a length of garden hose or conduit; or carried around and out of the building on ceramic insulators attached to structural elements and positioned well above head height.

The cases on most battery-charged models form weathertight containment for the controls and a 6-volt dry-cell battery. In my experience, these batteries seldom provide maximum charging power for more than 60 to 90 days. Such batteries are not all that costly, but you do pay for the convenience of a charger that can operate far from the nearest electrical outlet.

Old 6-volt car batteries that are still in good order can be used to propel these chargers. Each battery is generally positioned on a board on the ground beneath a charger; extra wire is used to reach from its post to the battery couplers inside the charger. There is even a new generation of battery chargers for short spans of wire that are powered with flashlight batteries and clip right to the fence wire.

The most expensive chargers are generally the solar-powered models. They can give dependable service even during moderate periods of cloudy weather and can carry electric fence usage to the farthest reaches of the family farm.

Electric fence chargers are one of the few things you can operate for pennies per day.

How to Ground Chargers Properly

Battery- and solar-powered chargers can be set right at the head of the fence line, but what is critical to the successful operation of all three types of electric fences is having the chargers properly grounded. A 6-foot-long copper rod driven at least 4 feet into the ground will serve admirably to ground a fence charger. The rod must be driven deep into the ground to ensure that it is constantly in contact with soil moisture. In some areas, a 6-foot rod may have to be driven nearly its full length into the ground. In very dry weather or climes you may find it necessary to regularly douse the area around the ground rod with several buckets of water.

Electric fencing materials. A number of fencing material options are available for use with modern electric fence chargers. There is a mesh material that resembles woven wire, but that can be folded and unfolded for rapid erection and take-down, ease of relocation, and simple storage. There are tapes and ribbons that combine plastics with fine wire fibers to create high-visibility fencing that can be spliced simply by tying ends together. There is even a light-gauge barbed wire that can be used with electric fence chargers.

The most often-used wire choice for electric fencing is 9- to 14-gauge smooth wire. It can be rolled and unrolled for repeated use, is durable, carries the electric charge well, and can be erected over virtually any terrain. It is commonly sold in spools that contain ¼ mile of wire each.

Posts for electric fencing include such options as hollow and solid fiberglass rods; short, smooth metal posts; and even homemade models that give old concrete rebar a second lease on life. Most are 42 to 60 inches in length and go into the ground easily with a few taps from a shop hammer. In many soil types they can even be pushed into the ground by hand or by simply stepping on their flanges.

The fence wire can actually be threaded through some of the fiberglass posts, or attached to them with simple metal clips that resemble cotter keys. They are ungrounded, and the current simply flows through them, too. To the wooden and steel posts, a variety of plastic and porcelain insulators can be affixed in various ways; these insulate the charged wires from shorting out against the posts.

With metal posts I prefer to use the plastic insulators that can be slid up and down without disconnecting the wire. This is an especially useful feature when electric fencing is being used to contain growing hogs, or in areas where substantial snow buildup occurs.

A single strand of charged wire suspended about 12 inches above the ground will contain most hogs from about 80 pounds up to breeding animals. At this height a hog will most often contact the charged wire across the bridge of its nose, a point that is quite effective in turning the animal back from the fence line. A second charged wire 4 inches above the ground will contain small pigs and young shoats.

Erecting the electric fence. When erecting electric fencing, I like to use 6- or 7-foot-long steel posts that are driven into the ground until the flange is well covered for my corner posts. I use porcelain doughnut insulators attached to the posts with heavy-gauge wire to anchor the line wires to the corners. Then, regardless of ultimate pen shape, I can run the wire in straight lines from one anchoring post to another. At the corners I join the line runs with short pieces of wire wrapped around the two lines in a coil-like manner. This ensures very taut fence lines and is very quick and simple to erect.

Donald Bixby/American Livestock Breeds Conservancy

A single strand of charged wire suspended about 12 inches above the ground will contain most hogs about 80 pounds and up. But for smaller pigs, a second charged wire about 4 inches above the ground will be needed.

Electric Fencing: Helpful Hints

I have used electric fencing for many years to contain a number of live-stock species. Along the way, I've picked up a few bits of lore and wisdom on how to operate and maintain the fence line and charger for best results:

◆ When charging multiple pens with a single charger, install simple, blade-type switches in the various pen lines to simplify repair work. Then you merely throw a switch to isolate a segment of the line or a pen to make it safe to repair or even dismantle an entire pen.

◆ Hogs that have never been exposed to charged wire tend to run right through it on their first encounter. To give them their very first experience with hot wire, expose them to it inside a woven-wire- or panel-enclosed lot. Run a short strand through the pen, place a bit of feed adjacent to and beneath it, and they will soon learn to respect it and give it a wide berth.

◆ Monitor the charger and fence line several times a day. Walking the lines and checking the charger while doing chores is a good practice. We've owned sows that knew immediately when fence chargers went off line.

◆ Walk the lines completely following strong winds and storms. Weeds, grass that's damp and heavy, and limbs across the fence line can short it out totally or greatly reduce its ability to contain hogs.

◆ Prefer solid-state fence controllers to those with replaceable, plug-type chopper controls. They cost more, but stay on line better, and you don't have the often-repeated cost of replacing the chopper control. Likewise, the fences with the ability to shock through wet weeds and grasses are preferable.

◆ Invest in one of the simple testing tools for measuring the strength of the charge being carried by the fence line.

◆ Hogs are often reluctant to cross areas where they know charged wire has been used, so be sure to install frequent gates or panels in electric fence lines. Where pens come together, set a panel in the line and then butt two more off it to form gates for the separate lots. You can then add additional panels in those corners to form working pens in those areas. In these small pens within bigger ones, the hogs will feel more secure and at ease, and they'll be easier to sort, handle, and load.

The Improved Range House Design

A recent, much-improved range house alternative is a simple rethinking of its basic design. Instead of a long side, it is one of the narrow ends that is left open to face south or east. The two runners remain the same length, but they are placed under the sidewalls rather than across the back and front of the house. Such a house can be easily towed into, through, and out of a pen.

Its real benefit, however, is the greater comfort quotient for the hogs it contains. With a 5-foot front opening falling to a 4-foot-high back wall over a length of 12 to 16 feet, a true micro-environment is created deep within the house. Bedding is kept dryer, chilling winds are less likely to penetrate to where the hogs lie, and body heat is better contained within such a house. Some ½-inch blackboard or other sheet insulation should be placed beneath the sheet metal roofing. This prevents the sleeping hogs' breath from condensing on it in cold weather, with the condensation then falling back on the hogs and bedding and chilling them.

For size and logistics of movement, few pieces of hog equipment are as daunting as a range house. I recall attending one farm dispersal sale where two really huge, 14- by 24-foot range houses were to be offered at auction. They would not fit on any of the commonly available trailers of that day — they had actually been built with no intention of ever removing them from that farm. They sold for less than many of the smaller houses offered that day, and the reason for that was best explained in the words of a farmer I stood next to in the crowd: "Ain't only one way to get them things home, boy. Just hook onto 'em and don't look back till ya get there," he said.

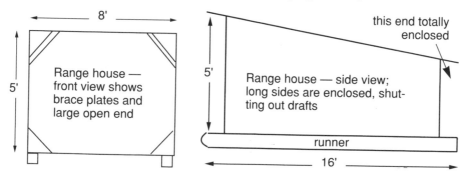

The newer type of range house has an improved design that features an opening on the narrow end, not the long.

The Range House

The basic unit for housing growing and breeding hogs is the range house. It's the Chevy of hog housing, and it can be used in drylots or pasture. Simply put, it is a three-sided house on runners with an insulated roof.

Basic Design

Generally, the range house has no floor, and varies in size from 8 by 10 feet up to 8 by 16 or larger. These houses are positioned with the open end facing south or east, away from prevailing winds in most of the United States.

The traditional approach to range house construction was to leave one of the wide sides open and extend the roof overhang 18 to 24 inches out over the opened side. The overhang was a house's only protection from blowing snow and rain. The design saved some building materials, and, with such a wide opening, the hogs entered and exited with little or no strain on doorways and structural elements. But that wide-open side was also an Achilles' heel.

Despite the overhangs, rain and snow could sometimes blow all the way to the back of the shallow beds. The house was more open to chilling winds, too, and in cold weather a large portion of the open side had to be boarded up to help contain bedding and body heat. Further, with its long open side this house was more cumbersome to position in or adjacent to a pen, especially a smaller pen. The improved design (see page 138) is superior.

Range House Interiors

Inside the range house, each sow will need at least 12 to 16 square feet of sleeping space; an older boar may need up to 20 square feet. In cold weather adding as little as 4 inches of clean straw bedding can raise comfort levels by an ambient level of as much as 10°F.

Range houses have a great many uses in the small-scale swine operation. One of our neighbors has a number of 8- by 16-foot houses he uses to shelter sows and litters while on range. At one to two weeks of age, depending upon the weather and the time of year, he moves nursing litters and sows from their farrowing huts to wooded lots or pastures. He puts two or three sows and litters in each house, shutting them inside it with a hog panel wired across the front, and holds them there for a night to get them accustomed to their new quarters. The sows and litters are maintained in these houses until weaning.

In many places, an 8- by 10-foot (or a bit larger) house is used to make a portable nursery to hold the pigs from several litters following weaning. A solid floor of 2-inch-thick native hardwood lumber or plywood is put into the house, and the front opening is reduced to a couple of small, pop-through doorways, which will admit only small amounts of potentially cold air to the sleeping area. The house is then pulled adjacent to a slatted-floored pen of equal dimensions. Feeding and watering equipment is set in the outside pen. Pigs of up to 40 pounds will need about 4 square feet of floor space each in such a unit.

Moving Large Hog Houses

There are a few simple measures you can take that will render larger hog houses easier and safer to move about. Some suggestions:

1. Bolt towing straps to both ends of each hog house runner. Use U-shaped straps at least 1 inch wide and ⅜ inch thick, and bolt them onto the runners through holes bored back at least 1 foot from the runner ends. Any shallower and there is a very real risk of tearing them out during a very hard pull.

2. Brace corners carefully with 2-inch-thick hardwood lumber at floor level, and plywood plates at the roofline.

3. Start a house moving with a slow and steady pull, not with a jerk.

4. Set the house on a 4-inch-high raised bed of tamped earth, cull lime, or gravel. This will keep runoff from passing through the house and the runners from settling deeply into the mud.

5. A really old trick is to set each corner of the house down on a large fieldstone to keep the house from settling into the mud, making it harder to move by towing.

6. Trench around the house to further direct runoff water away from the sleeping beds; keep the trenches cleaned out to prevent mud from building up around the runners.

Keep in mind that the longer a house sits in one place, the harder it will be to free up and move.

Farrowing Structures

The other big structural requirement for those with a sow herd is a place for farrowing — somewhere to shelter the females during birth and for a few days afterward, until the pigs are off to a good start. In many months of the year a sow can heap up a pile of leaves and old cornstalks and raise at least a few pigs, but if you have litters coming at every season you need housing that shelters the newborns from the elements, predators, and muck and mire.

Choices in farrowing housing will be dictated by economics, climate, and enterprise goals. The producer selling boars and show pigs must be able to farrow in January come what may. The producer wanting to sell F_1 gilts or butcher hogs may be better served by a more seasonal approach to farrowing and the resulting cost savings. On some family farms, a combination of these options may be in order.

Farrowing house options range from simple floorless huts and hutches, to one-sow houses with porch pens, to pull-together or modular houses, to centrally located, single-purpose farrowing houses. All work well in specific situations, and there is something to be said for and against each and every one.

At their core is the means by which the sow is contained during the time of farrowing and early pig rearing. Sows can be contained in farrowing crates, stalls, or single-sow houses (the latter is basically a four-sided version of the freestanding farrowing stall).

Farrowing Crates

Even after decades of use, farrowing crates remain somewhat controversial, due to the constrictive way in which they contain sows. Crates do indeed hold the sows in very close quarters — but for the express purpose of protecting very young pigs from overlay and crushing. During the first 96 hours following farrowing, death losses are greater than at any other time in the pig's life, and overlay is one of the primary causes of those deaths.

The basic crate is 5 feet wide by 7 feet long. The central part of the unit, where the sow is contained, is 2 feet by 7 feet. It is supposed to be snug enough to keep her from turning around and stepping on any of her baby pigs, or flopping down on them when she lies down to nurse.

It holds the female in a rather exacting position, so that even the birth will occur in a specific portion of the crate. There the pigs will be easily accessible to the producer, supplemental heat can be directed upon them, and they can easily find the sow's side and quickly begin nursing.

Portable Farrowing House

The illustrations below show the framing on a portable farrowing house developed by extension personnel for LaClede County, University of Missouri — Columbia. It features a sliding roof, and it can be moved by a forklift on the back of a tractor. The forks slip between the skids, and the house is lifted by a hydraulic system. The house also can be pulled on skids, although in this case the skids should be sloped at the ends, similarly to skids for sleds. Notice that one side is for pigs and includes a pig guard, while the other side is for the sow.

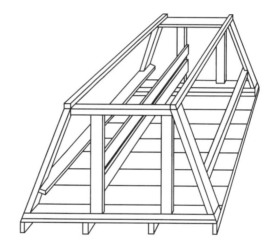

The floor and framing for a LaClede County house.

An enclosed house showing boards in place to reduce overlaying by the sow on young pigs.

Down each side of the sow's zone are 18-inch-wide by 7-foot-long pig bunks. Here the pigs can lie safely away from the sow, still have simple access to her side for nursing, are safely provided with supplemental heat, and can be proffered creep feed or milk replacer. These bunks are meant to be a safe retreat for the pigs away from the danger of overlay or misdirected maternal feet.

Initially, farrowing crates were made of 2-inch-thick planking arranged in a gatelike fashion with an 8-inch-high opening at the bottom for the pigs to have easy access to the sow when they wished to nurse. This crate design is still used in many areas due to its simplicity and modest cost; it can also be homemade with simple tools. However, over the years a number of modifications have been made to this design, and the great majority of crates are now made of steel pipe or square tubing.

Simple farrowing crates are built into some one-sow farrowing houses, also. A midwestern variation on the farrowing crate has metal-framed crates erected on 5- by 7-foot sheets of expanded metal flooring set on steel pipe legs. The crate frames, including the pig bunks, are enclosed with plywood sheathing. They can be set up in various outbuildings for cold-weather farrowing, or taken to the hedgerows and field margins for summer farrowing. Wastes fall through the mesh floors, and the crates are fairly simple to take down for cleaning or storage.

Kelly Klober

The great majority of farrowing crates are now made of steel pipe or square tubing.

A narrower crate — 22 inches wide in the sow area — is often used to farrow gilts, which are both younger and smaller than sows. Another modification if space is at a premium is to provide a pig bunk down just one side of the farrowing crate. Some producers also install a nest box at one end of the farrowing crate to hold young pigs completely away from the sow's zone in a micro-environment all their own.

Crates do save young pigs, and with proper management they can maintain sows in a comfortable and healthful manner. We have used crates made from native oak lumber; while deaths from overlay and crushing did go down, we also lost a few pigs to cuts caused by sows extending their hooves into the pig bunks.

Newer Crate Options

Among the most recent modifications to the basic farrowing crate are steel fingers and a hydraulic action on the bottom crate member along each side of the sow. As the sow begins to settle down, she is discouraged from dropping down on any pigs in her zone, and the rotating fingers push the pigs away from the sow. Another modification even uses puffs of air to move pigs away from a descending sow. Such state-of-the-art features add substantially to the costs of a farrowing crate, so many of them are best suited for the newer versions of the environmentally controlled farrowing units.

Crate flooring. Nearly as oft-discussed as crate design options is the flooring material that goes beneath the crates. On most small farms, crate flooring is still found in its most basic form — some sort of solid wooden flooring to keep drafts from coming up under the sow and her litter and chilling them. For us, native oak 2 x 6s or 2 x 8s laid side-by-side without slots provide both draft control and durability and are among the least-costly flooring options.

There can be problems with wooden flooring under hogs, however. When wet or covered with manure, it can become quite slick. Plywood is an especially poor choice for flooring beneath hogs of nearly any age. Treated wood can sometimes burn sow udders and the delicate skin of very young pigs; white-skinned pigs seem especially vulnerable to such irritation. Also, very rough flooring sometimes causes a problem with navel ill. This is an infection of the umbilical cord stump that you'll learn more about in chapter 9, but it can begin after baby pigs' delicate underlines are irritated by lying upon or squirming across rough wooden surfaces. Navel ill can sometimes even go on to result in umbilical hernias; little gilts can suffer irreversible damage to their teats.

I once had a problem with navel ill that started when pigs crawled over a 2 x 4 nailed to the pen floor to support a pig bunk partition. We removed the offending board and the problem was quickly eliminated, but we remain sensitive to what happens in the little pigs' environment, which is so easy to overlook when going about chores and seeing and experiencing things only at head and shoulder level.

Over time I have seen folks fasten rubber pads to the floors beneath sows and pigs as well as trying combinations of slats and solid flooring, various types of expanded metal and mesh floorings, and even combinations of wood and mesh. In vogue right now is the practice of elevating the whole crate 8 to 10 inches above existing solid floors, placing 5- by 7-foot wire mesh pads beneath the crates, and letting wastes fall into shallow pits also set beneath the crates.

Normally, sows go into farrowing crates three to four days ahead of their due date to become accustomed to and at ease in their new surroundings. Following farrowing, they may remain in the crates anywhere from a few days to until the pigs are weaned, at between 14 and 56 days of age.

Some additional points on farrowing crate management and use include:

1. The sows should be treated for both internal and external parasites shortly before being placed in the crates. Some producers give the females a good scrubbing down with mild soap and water first, and some have even devised special "sow showers" for use before placing them in the farrowing crates.

2. The sows can be fed and watered in the crates, but the equipment to do so properly can be rather costly. Also, anytime water goes into the crate, problems with dampness and slick spots in the floor can occur.

3. In many smaller herds the crated sows are turned out twice a day, night and morning, for about 30 minutes, to eat, drink, and stretch their legs. Many sows will even wait until they are out of the crates to dung and urinate. With this system, the sows are generally kept more comfortable, and while they are outside you can care for pigs and clean or rebed the crates.

4. In cold weather I always lay sheet tin directly atop the farrowing crates and use straw bedding to help hold in body heat.

Farrowing Stalls

Farrowing stalls are an old, old farrowing option that still has a lot of adherents. Such stalls are generally 4 to 5 feet wide and 10 to 14 feet long. The back 2 to 3 feet of each pen is partitioned off for baby pig bunks. We fitted six such 5- by 12-foot stalls into one bay of our old Missouri horse barn, sealed it up with plywood sheets on hinges that could be raised to increase air circulation in hot weather, and farrowed in it quite comfortably 12 months of the year.

Stalls give animals a great deal of freedom of movement and can contain both sow and litter easily until weaning. Pig losses to overlay will be a bit greater — although 2 x 4s attached on edge down each side of the pen will do a surprisingly good job of protecting very young pigs from being crushed as the sows lie down to nurse.

Also, if allowed too much straw or other bedding material, a sow might build a large, mound-type bed or nest on which to farrow. If she farrows while lying atop such a mound, her newborn pigs may not be able to reach her side to nurse. Such pens will also require extra labor to keep clean and dry.

The Pull-Together

A third option is to incorporate crates or stalls into a modular-type building called a pull-together. This is normally built in two halves, with a length for each half of 10 to 30 feet, a width of 7 to 8 feet, and an additional roof overhang of 3 to 4 feet; the two halves can then be pulled together to form a modular farrowing building with a large central alleyway.

Depending upon its size, the house may contain 4 crates or more, or stall up to 10 or 12. The larger halves may require a tractor of at least 50 hp to tow them about and position them. This is, however, a housing option with many virtues. Consider the following:

1. Per crate or stall, the pull-together is one of the least costly of all farrowing options. Our first 8-crate house cost just $650, and a 10-crate house can now can be bought new for less than $3,000. A used house can be found for just about a third of that.
2. The house can be easily taken apart to be used throughout a pasture rotation system, and as a further sanitation measure.
3. It is of a very handy size, but does not appear on property tax rolls as a permanent structure.

4. For income tax purposes it will often qualify as a depreciable, special-purpose expense. Its cost may even be entirely deductible in the year of purchase in many circumstances.
5. It can be moved from farm to farm should you ever decide to relocate.
6. It is suitable for farrowing 12 months out of the year.
7. With only minor reworking it can be made usable for other species. It can be used for lambing, rearing bucket calves, or even housing poultry.

Single-Sow Units

The option I favor here at Willow Valley is the single-sow-and-litter farrowing unit. In many parts of the country the solid-floored version of this 6-by 8-foot house is called a "Smidley," a trade name taken from the Ohio firm that makes the distinctively orange-colored farrowing units as well as other livestock equipment. In this same vein there are simpler still three-sided, floorless huts that are used for seasonal pasture farrowing.

My choice for year-round farrowing is a row of 6- by 8-foot farrowing houses fronted by slatted-floored pens of the same dimensions as the house and then set 8 feet apart. That 8-foot spacing is adequate to prevent the aerosol spread of disease organisms.

The houses can either be set up as 6- by 8-foot self-contained farrowing stalls, or farrowing crates can be built into them with 2-inch planking. With either option, 2- by 4-inch guardrails should be installed down each sidewall 8 inches above the floor.

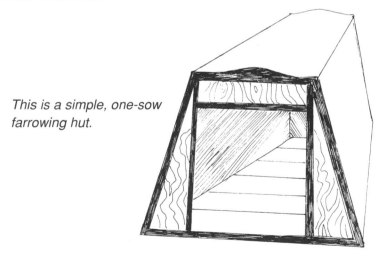

This is a simple, one-sow farrowing hut.

I favor a house with a walk-in door on each end. The doors provide good access to the sow and litter, especially through the rear door into the pig bunk. Roofs can either slide back or lift up, which is useful in regulating air flow across the animals in hot weather, and makes cleaning easier.

The doors are generally 18 to 20 inches wide and 30 to 36 inches high. A piece of a 34-inch-high hog panel cut just a bit wider than the rear door opening can be fitted across it in warm weather; then you can leave the rear door open to further improve air flow. And in winter here in east-central Missouri, these houses do an excellent job of sheltering a sow and litter with the addition of just one or two 125-watt heat lamps. In a tightly constructed house with both doors shut, a sow and her litter will generate as much as 6,000 Btu's of moist heat per hour.

Most three-sided huts are 4½ or 5 feet wide by 7 feet long. They are generally made entirely of sheet metal on a lightweight metal frame and resemble small Quonset huts, or are of corrugated sheet metal on a wooden frame. Such houses are often so light that one person can load and unload several of them from a pickup — even flip them up and drag them about for positioning in lots or on pasture. If not adequately staked down, some of the lower-height models can actually be moved around by big sows in a turtle-shell manner. Seeing a sow walking across the pasture wearing her house will certainly bring you up from the supper table with a bit of a start.

The sows need to go into the houses early enough for each sow to select her own house and stake out her own territory — at least three to four days before the first sow is due to farrow.

Seasonal Farrowing on the Range

Seasonal farrowing on range is possible six to eight months of the year in most parts of the country, and even is pursued into Michigan and Minnesota. On pasture the houses need to be positioned 100 to 150 feet apart, to keep sows from doubling up in them and thus increasing pig losses through crushing.

Housing for Sows with Nursing Pigs

Some producers set their houses in place first and then drop off a sow at each house site. The huts for range farrowing need some sort of bar or stop across the doorways to prevent the little pigs from leaving the safety of the huts until they are at least several days old and can navigate the nearby terrain and safely return to their own beds.

Long rollers can be attached across the bottoms of the doorways for this purpose; a newer innovation is a small enclosure 8 to 10 inches high that encompasses a few square feet in front of the hut. There, pigs can venture out a bit while still being safely contained, and the sows can simply step over the low enclosures to venture farther afield.

In very cold weather, the Smidley-type houses described under "Single-Sow Units" (see page 147) can be drawn inside larger structures such as barns or machine sheds. There, the animals are protected from strong winds and intense cold, electricity may be available for supplemental heating, and the houses can be drawn closer to home and feed and water supplies.

Readying Supplemental Heat Sources

You should have a supplemental heat source for young pigs ready to go before farrowing. Keep in mind that sows can be bred to farrow 12 months out of the year. There are a number of ways to provide the heat very young pigs need to stay safe and warm. The trick is to warm the pigs without causing heat stress to their mother.

Among the heating options are forced-air units with the capacity for heating whole buildings, flameless gas heaters that use ceramic heating heads, electric heating mats that warm the pigs from below, and electric heat lamps. Wood heat is also often used.

One Amish hog raiser of our acquaintance uses wood heat supplied in a novel way to heat his farrowing house. The firebox sits outside the farrowing house on one of its ends, with the flue on the other. Heat and smoke are thus drawn through a large pipe in the farrowing house floor. The heat radiates upward through the concrete floor to warm the young pigs at the back of the farrowing stalls. It provides a fairly efficient comfort zone for the young animals.

Trying to heat a whole building to the temperature at which very young pigs are most comfortable can be quite expensive. It can also cause heat stress to the sows, with a resultant decline in milk production and restless behavior. A sow is most comfortable at 55 to 60°F, whereas a newborn pig has just left an environment with a 90°-plus temperature. The best compromise is to create a zone in which the pigs can be kept comfortable without overtaxing the sow or the pocketbook.

Creating A Comfort Zone for Baby Pigs

We rely greatly on heat lamps and pig bunks at Willow Valley, but there are hog producers who use all kinds of combinations of the following. Many of them "tinker" until they find the combination that works best for them. Here are some of your options:

◆ Hovers, which are three-sided boxlike enclosures that fit in and over a portion of the pig bunk and are accessible only to the pigs through small "pop" holes, create a snug, easy-to-heat zone. Made of plywood, they simply pull in and out of the pig bunk. Some fit over a heat pad; with others, an electric heat lamp shines down through a circular hole cut in the top.

◆ Heat pads are made of a sort of armored plate or fiber, attached securely to the floor and positioned well away from the sow. They are regulated with a thermostat, and the pigs are warmed by lying atop. For a few days after farrowing you may have to use a heat lamp to draw the little pigs to the area warmed with the mat.

◆ Red-tinted heat bulbs seem to do a better job of drawing the pigs to hovers and bunks than clear or white-frosted bulbs.

◆ Insulating material can be taken right down to floor level in farrowing quarters if it is protected from damage by the sow and pigs. Covering the insulation with l- by 2-inch wire mesh will shield it from the hogs.

◆ Even one-sow houses can be tightened and made snugger in cold weather. An 8- by 10-foot pickup tarp drawn tightly around the end of the house that holds the pig bunk will greatly reduce drafts. A few bales of straw across the bunk and up the house sides will further insulate the house. One very cold and snowy Missouri winter I recall seeing neighbors heaping snow around and over houses in somewhat of an igloo fashion.

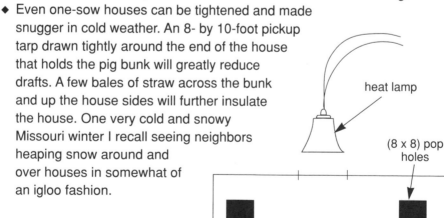

Safe Use of Electric Heat Lamps

On most small farms the primary utensil for providing supplemental heat in any season is an electric heat lamp with a metal reflector. It is lightweight, inexpensive, simple to use, and can burn down a barn in a New York minute if not managed properly.

Steps to follow to use electric heat lamps safely include:

1. Never use more than seven of the 250-watt heat bulbs on a single circuit. The 125-watt bulbs will do a good job in many months of the year and will cost measurably less to operate.
2. Never suspend the lamps by their electric cords or fasten them up with string or baling twine. You can use smooth wire to suspend the lamps, but the safest choice is lightweight chain.
3. Never suspend the lamps lower than 24 inches above the pigs. The lamps warm by heating what they shine upon and thus can badly burn tender-skinned young pigs, or even ignite bedding.
4. Be sure to keep raising the lamps as the pigs grow. A pig which can rear up to a rather surprising height when even a few days old, can pull down lights or cause bulbs to burst by touching a burning bulb with the tip of its nose.
5. The lamps must also be suspended in a position well away from the reach of the sow. Sows will rear up and climb on gates or crate rails to reach heat lamps.
6. Check the lamps often while in use for frayed cords, dust buildup on the reflectors, any slippage out of position, and loose bulbs.
7. Steadily raise the lamps to wean the pigs from their dependency upon the extra heat. Seldom do I use them for more than 14 days (generally 7–10 days). We often replace 250-watt bulbs with 125-watt within 72 hours of farrowing.
8. In extreme weather we have used two lamps to get a young litter through a cold night. In such conditions, a second lamp suspended above the sow's hindquarters at farrowing can increase pig survival. It must be monitored quite closely, however.
9. If outside air temperatures are above 60°F, you may not need supplemental heat.

Seven

SELECTING BREEDING STOCK

TO SELECT BREEDING STOCK you build upon the principles of animal selection outlined in chapter 2, but you must also factor in the type traits and genotype that influence reproductive performance. Of all of the economic and carcass traits you will hear about when discussing swine selection, never forget that none are more important than live young. To get these you have to have animals with the will, vigor, and conformation to do those most basic of tasks: live and reproduce.

Genotype is the two dollar word for the inward genetic makeup of an animal. It is what the animal can do for you in such areas as growth rate and carcass type. *Phenotype* is how the animal's genetic makeup manifests itself visually. It covers traits such as conformation, soundness, and muscling that are at least somewhat apparent to the naked eye.

There is no more important or better place to spend money than on the acquisition of quality breeding stock. In breeding stock you do indeed get what you pay for — but if you shop carefully and invest wisely you actually get a bit more.

Shopping for Livestock

A few years back the pork market was going through one of its boom periods, when anything 250 pounds or heavier and female sold like it was gold plated. Being a firm believer that the time to start or add to a livestock venture is when the desire and the dollars come together, I was in the market for a couple of extra gilts.

Down at the sale barn, gilts were busting butcher hog prices with $100 to spare, the seedstock companies were making car dealers look like pikers in their advertising, and breeder auctions were followed immediately by trips to town to see if Rolls-Royce made pickups. Yet in just 20 minutes on the phone I found a set of gilts less than 30 miles from home that were priced only $75 per head over market, had good pedigrees, and were guaranteed to breed and settle.

I've found the best places to find good buys on hogs are (in order of preference) talking with fellow raisers, attending hog events, reading local newspapers, and checking out hog publications.

To eliminate the disease pseudorabies, far fewer sale barns and community auctions now sell breeding animals, especially not breeding boars. I think this is a good thing, because a lot of folks were buying boars there that weren't paying off. Most boars today are purchased from private breeders.

Swine production is one of the few ventures in which you can get on the phone and order up hogs sight unseen knowing that they are backed by a code of fair play adopted and sanctioned all across the industry. This is a semiofficial code adopted by a number of breed associations with the basic element, of course, that the hog can be returned if the buyer is not satisfied. It sure does make seedstock shopping a whole big bunch easier. Still, there are a few things that the savvy shopper does need to do. Here are eight tips:

1. Compile a list of prospective suppliers, contact them first by phone, and make appointments to go view hogs if they sound promising. It's just not fair to roll into a driveway and expect a producer to turn off a tractor at the height of planting or harvest to show you hogs.

2. Start your seedstock shopping well in advance of need. Going to look for a boar when the gilts back home are already going through their third heat cycle marks you as a real rube. It can also get you saddled with a sale barn sow settler — an animal that just won't advance the herd — rather than a boar that will move your venture forward.

3. Select seedstock with the idea of improving or upgrading no more than one or two traits at a time. Try to key on the most glaring fault in your sows or their offspring and work at improving it.

4. Buy from facilities as similar to your own as possible. This helps minimize stress on the animals.

5. On the seller's farm, try to view siblings, sire, and dam to ascertain just how dependable the genetics really are. Are they truly breeding on and breeding "deep" — that is, are they producing quality animals in good numbers?

6. On the seller's farm, don't hesitate to ask questions on everything from how the hogs were bred to how they were fed. A good animal might be redeemed from neglect or poor nutrition, but not if it has been stunted; nor will the best of care overcome poor genetics.

7. Don't haggle over price. If it's too high, thank the producer kindly for his or her time and look on down the road. That way you can both feel comfortable about possibly doing business together in the future. Also, be realistic. If your budget is tight, don't go looking at state-fair winners.

8. Form a mental image of the kind of hogs you want and carry it with you into the lots and pastures that you visit. Be fair, however, and realize that the perfect hog has never been bred. It is perhaps best to go with first impressions when selecting breeding animals.

Performance Data

When you're shopping for breeding stock today you can literally encounter a deluge of statistical data. Through various scanning and probing devices, many of them ultrasonic, it is now possible to gauge loineye area, fat cover, and muscle mass on a living animal without harming it in any way. Actual slaughter data may be available on part and full siblings. Performance data may also be available on one or both of the animal's parents, and on some proven breeders there is sufficient collected data to estimate a bloodline's influence on a number of specific traits.

Many years ago, when the trend toward collecting performance data began, a set of minimums for boar performance was established and almost codified. A boar being considered for service had to have at least a 4.5-square-inch loineye, a maximum of 1.5 inches of backfat depth, and maximum days of 160 to 220 pounds in weight. There have been many changes since this early set of standards, though, and nothing quite so constant is now in place.

Days are now measured to 230 pounds or even heavier, backfat thicknesses have slipped to 1 inch and often much less, loineyes on boars now have 5.5 square inches as a minimum (6-plus square inches is now quite common), and butcher hogs regularly top 70 percent lean yield. Hogs tested for growth and feed efficiency (such tests are conducted by individuals, universities, and breed groups) were once producing 1 pound of gain on 3.5 to 4.5 pounds of feed. Now that figure is closer to 3.5 pounds, including the feed consumed by the hog's sire and dam during breeding and gestation. Also, tested boars have of late begun closing in on the 2:1 feed efficiency ratio once thought possible only with broilers.

Minimum Boar-Performance Standards

The minimum performance standards for boars have changed considerably over the years. Following are old standards compared with the widely accepted standards of today.

Performance Factor	Old Standard	Today's Standard
Days to market	160	160–175 (leaner hogs grow slower)
Market weight	220	230
Loineye	4.5-sq.-in. minimum	5.5-sq.-in. minimum
Backfat	1.5" maximum	1" maximum
Lean yield	there wasn't one!	70% or more
Feed:gain ratio	4.5:1	3.0–3.5:1

Indexing Pitfalls

Performance testing refers to individual data on each animal per specific trait and is the most reliable method of assessing that animal. Beware of data presented in any other way — at least to an extent. For example, information is sometimes presented in the form of indexes that give a numerical score, over or under the herd average, for performance of a particular trait. The average is often given a numerical value of 100; better performers will score above 100, and below-average performers will receive a score below. Seldom will scores move more than 5 to 10 points above or below that average figure.

I find indexes to be a bit deceptive. A boar scoring a 98 for growth is measurably below the herd average, but many of us with a public school education remember 98 as being a quite good score. A score above 100 denotes performance above the herd average, but what if the average wasn't all that good?

The current trend toward exceptionally lean hogs and the corresponding slowing in growth rates has a lot of sellers opting for indexes to score growth traits. Some exceptionally lean hogs may take as many as 10 to 20 extra days to reach good market weight.

Changes in statistical goals and uses have not been without other consequences, either. With the leaner hogs not only are growth curves slowed, but traits such as litter size and reproductive performance can also be demonstrated to have declined. Further, there are several very important traits that cannot be denoted statistically.

Misleading Figures

I can recall a very noted Duroc boar of a few years back that quite literally tore up a major midwestern boar test and rewrote the record books. His offspring, however, failed to perform up to expectations, were lacking in desirable conformation traits and good breed character, and often had soundness problems. At one of the public displays of this boar at a breeder's auction a veteran farmer sitting next to me in the stands summed up the animal and his value quite succinctly. He said, "The day they went to buy that boar they should have spent more time looking at him and less time reading about his figures in the sale catalog."

Weighing the Data

Performance data is an important selection tool, and, while estimated breeding values or performance differences are still very new and much debated, they are dramatically building the sheer numbers that can now be presented with some hogs. All of these numbers need to be weighed very, very carefully, however.

The way data is obtained and released is still very much under the control of individual breeders, breed associations, and test stations. The performance of young boars is enhanced by the presence of male hormones that will be unavailable to barrow or gilt offspring. And some data is obtained in ways that are neither realistic nor practical for most family farms. Test groups are often quite small and are often full siblings; they're quite closely confined and often fed rich, rather costly and complex rations to maximize performance. They are burning high octane to develop data that is supposed to be used in a real world run on regular gas.

It should also be borne in mind that the traits for which most of the data is available are the traits with the highest degrees of genetic inheritability. Those with a lesser degree of genetic inheritability are no less important to the overall success of a swine operation, however. Litter size, for example, has a quite low degree of genetic inheritability (it is a trait as much or more influenced by nutrition and environment), but if we let it slide in our Duroc herd it soon hits us in the pocketbook and hurts our boar sales.

We'll get into more particulars later, but other factors that must be weighed in seedstock selection include litter size at both farrowing and weaning, length and body capacity, foot and leg structure, the development of reproductive organs, muscle pattern, frame size, sexual character, and more. To determine these points you have to work the hogs (move through them), watch them move about, view their home environment, and learn something of their history.

I began my hog-raising career with a sow about halfway through her productive life, but it is far more common to begin with open or bred gilts. *Open* is a producer's term for a female of breeding age that isn't bred. Some gilts also are sold as exposed or pasture-exposed to a boar, but there's no assurance that they are pregnant. Gilts will always be a bit of an unproved commodity, but a long, productive life should lie before them. It is easier to determine if a young female has merit as a producer of meat animals than it is to tell if a sow has this potential; pregnancy and nursing alter a female's appearance and you won't find the tightness and definition of muscling in a sow that you will in a young gilt.

Gilt Selection

The function of the gilt is to bear and mother the young, but this does not mean that you can neglect growth and meat type during the gilt selection process. At one time it was a fairly common practice to pull replacement gilts from the last hogs remaining in the finishing pen. This was actually selection for slow growth and weak type.

Home-Raised Gilts

You should select the biggest, most well-grown and -developed gilts in the group, even if it delays a payday for a week or two. If you are building a herd with home-raised gilts, initial selection should come from the top half of the early pig crops. In time, the quest for quality should surpass the need for numbers, and only the top 10 percent of the gilts produced will be good candidates to enter your breeding herd.

In a well-managed herd, sows should be regularly producing gilts that are better than themselves, which should further facilitate gilt selection. As an average, most swine producers turn something on the order of 40 percent of their sow herd yearly.

In the selection of home-raised gilts the first cut is generally made on the day the pigs are born. Keeper gilts should only come from gilt litters in which at least eight large and even pigs were farrowed. In sow litters at least nine pigs should be born into any litter from which gilts will be selected.

Some hog producers balk at selecting breeding animals from gilt litters. The first reason for this is that the gilts have no proven track record as to reproductive performance. Also, due to their smaller size, gilts are more likely to have been bred to younger, equally unproven boars. Older sows have had a longer time on the farm and thus a longer exposure to the organism population, or "bug soup," that is to be found on all livestock acreage. Through this exposure they have built up a stronger natural immunity to those organisms — one that they can and do pass on to their offspring.

Poland Breed Association

This is an excellent Poland China gilt. Notice her good top, substantial bone, and well-formed underline.

Why Home-Raised Gilts?

I am a firm believer in the home-raised gilt (and you will find that I have focused this chapter accordingly). A gilt's lineage, her temperament, and even the quirks of her dam and granddam should be fully known to the producer. You can also grow and develop a young female specifically for a long and productive life in the home farm breeding pens, which are like no others. Simply put, sow herd quality control can begin the day the gilt pigs are born.

Further, I believe that economics still favor the home-raised gilt for small and medium-size producers. Assembling a group of 5 to 10 gilts good enough to replace a breeding group in a well-managed small operation is neither simple nor inexpensive. The goal with herd replacements is not to equal, but to improve, performance, and to ferret out and then purchase the cream of the cream of the crop, which can be both time-consuming and costly. If you're doing your job right, you should find plenty of good ones inside your own fences.

Select Larger Pigs

It is a simple matter to take very young pigs in hand and give them a thorough examination. The best candidates for survival and long, productive lives are pigs that weigh 3 pounds or more at birth. I make special note of gilt pigs with a birthweight of 4 pounds or more; we have even had some pigs crowd 6 pounds at birth. By always selecting for larger gilt size over a period of time, you will emphasize the natural growth and female capacity that should also ultimately result in fewer farrowing problems and larger, more vigorous pigs at birth.

Underlines

Very early in gilts' lives is also a good time to check out their underlines. To be considered as a female herd replacement a gilt should have at least 12 evenly spaced and well-formed teats. Fourteen is even better, and uneven counts of 13 or 15 are not to be discounted, although there should be a minimum of 6 teats on each side of the underline. Do not select gilts with unevenly spaced, inverted, or pin (too-small) nipples.

Determine Growth Rate

The next cut in a home-raised gilt's life comes at about 150 pounds. First determinations about growth rate and efficiency of performance can be made then. Some producers will cut out gilts from the finishing group at this point, but I prefer to leave them on a good 15 percent crude protein growing ration and with their peer group, to see if they continue to perform up to expectations.

The goal is to have a gilt weighing 300 pounds by eight months of age, the age at which she should be bred if she is to farrow at about one year. At about 220 to 250 pounds keeper gilts should be sorted out and placed to themselves, pulled off the self-feeder, and then developed steadily to that 300-pound goal. At each cut the gilts need to be evaluated for soundness, continued rapid and efficient growth, continued feminine character, and meat type. Even to 300 pounds, gilts culled from the keeper pool will retain good value as butcher stock.

Backfat

At five to six months of age, and 220 to 250 pounds, the gilts can be evaluated with one of the many scanning machines and/or probed for backfat. Most land grant colleges offer these services through their extension offices, and a call to the local county agent should produce a phone number for these tests.

Backfat Probing

You can learn to do backfat probing yourself. It requires only a good catch crate to securely hold the animal in position, a scalpel, and a thin metal ruler with a sliding gauge. Standard procedure is to make three probes along the animal's topline and average them to develop an average backfat thickness determination, a good guide to the animal's overall fat cover. A simple procedure that will work for those selecting gilts for home use is to probe just once in the third position, farthest back on the hog, and add 0.1 inch to that figure. It results in an approximation, but still a good, usable figure.

These days it is generally possible to adjust performance data backward or forward to a standard of, say, 230 pounds. This allows you to compare the performance of a number of hogs of different weights and ages to a true constant.

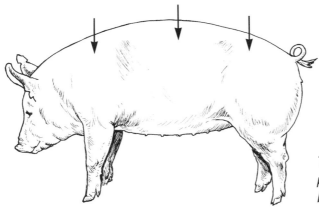

The three sites you can probe to determine backfat.

Conformation and Appearance

A good gilt will have a wide, deep body reflecting a capacity not just to grow but also to carry a large litter of pigs safely to term. Currently in the United States, most females don't make it past their fourth parity, but on small farms sows tend to hang in longer and so they have to have what it takes to endure.

Long, sloping pasterns, a well-formed leg set correctly on each corner, evenly sized and shaped toes, and a long and level top all contribute to long-term, continuing soundness. Large, well-formed bones as indicated by leg diameter are further indications of continuing soundness.

It is almost too abstract a term to explain, but a good gilt does have a truly feminine appearance. She must at once be large and well grown, yet retain elements of refinement and smoothness. Overall, a gilt will lack the stature and projection of masculinity of a male, will show refinement in and around her head, and will project a feminine image. Since our bread and butter is breeding boars, from time to time I have tried to use a large, coarse-headed female to create more masculine-appearing boars. Sometimes it works — but with well-formed, feminine-appearing gilts and sows, I get as many or more keeper boars in more even numbers and litters.

There is no device that allows the prospective buyer to peer into a female hog and view her reproductive tract, to evaluate her ability to cycle, breed, and carry large litters of pigs. The one visual clue to the capacity of the reproductive tract is the vulva, the only external organ of the reproductive system.

Duroc Breed Association

This Duroc gilt typifies what I mean by a refined, feminine appearance.

It should be large and well formed. Very small vulvas are the one visual indication that the rest of the reproductive system might not be adequate.

Balance Is Best

Gilts too extreme in type can sometimes wind up sacrificing performance in the breeding pen or farrowing unit. For example, gilts too trim along the bottom line and that fail to carry depth of side throughout the full length of the body may develop what is termed a "meat-type udder." Quite simply, the udder does not fill correctly, and one or two teat segments on each side of the udder cannot be used by the nursing pigs.

In selecting gilts, try to avoid such extremes in type. Go with what most call the middle-of-the-road type, which balances reproductive type with at least moderately good meat type.

Boar Selection

There is no faster way to bring about change or introduce new blood into a swine herd than through the purchase of a first-rate young boar. Something akin to heterosis has been noted in even purebred herds when a new boar of separate and distinct breeding is introduced. It is also the least costly and complicated way in which to bring new genetics to the small farm.

As I've explained, boars usually are bought from breeders, and bringing in a good one is a great way to improve your breeding herd. Actually, two good boars used in succession can effect as much as an 85 percent increase in the performance of a herd's progeny for a great many important traits.

It is also widely held that the real benefits of a good boar are not seen until his daughters enter the breeding herd and begin producing offspring of their own. This points up just how much you have to be concerned about when seeking out a good boar to head up even the smallest of herds. His influence, good or bad, will be felt for generations.

I have been to many breeder auctions at which the breeder, auctioneer, and breed association fieldmen worked long hours to select the right boar with which to start the sale and, presumably, bring top dollar and set the tone for the entire event — only to see that animal outsold by one appearing many places farther down in the sale order. Either the buyers don't agree with their assessment or the buyers are in the market for a boar with particular traits that the lead boar doesn't have. It all depends on who's shopping for what that day.

An old chestnut of wisdom in hog circles holds that the boars that make the big splash at sales or shows seldom produce sons and daughters that go on to make as big a splash. Their impact isn't generally felt until the second or even third generation, and a lot of big boars are never heard from again after their moment in the winners' circle.

Still, selecting a herd boar isn't like having a top-10 choice in the pro football draft. You don't go after the biggest gun available; you go for the boar that will solve a specific problem — your most pressing problem.

There is a two-fold task in selecting a herd boar: *Bring in the genetic piece or two needed to strengthen the one or two breeding weaknesses most common in your sow herd or their offspring, and don't lose ground in any of the other traits.*

As noted, most boars today are purchased from private breeders for about $300 to $400. They usually come with some sort of warranty or guarantee of soundness and breeding performance. You'll read more about that under "Boar Warranties" (see page 169).

Terminal Boars

A category of breeding boars that has only recently emerged is termed terminal boars. As I explained in chapter 1, *terminal* refers to a hog used as a cross to maximize growth and muscling, since the pigs produced are intended to be sold for butcher stock. This term certainly points up the role a boar plays in imparting the economic traits — growth, carcass qualities — to the offspring of a mating.

Terminal boars are generally F_1 crosses of two of the more noted meat-type breeds — Duroc, Hampshire, Berkshire, Black Poland, and so on — that are used to impart an approximate double dip of carcass and/or growth traits to pig crops. An interesting cross along these lines that is underutilized is the Hampshire x Spotted.

Boar Age

Most boars are bought as untried youngsters at 8 to 10 months of age, and weigh between 300 and 400 pounds. Some cash savings can result from buying a young boar 60 days or so away from service age (not a bad idea, because any breeding hog you purchase should be kept on your farm and in isolation for at least 30 days before coming in contact with the breeding herd).

Maternal Background

As noted, a boar's greatest impact often comes when his daughters and granddaughters enter the breeding herd. You must therefore carefully weigh a potential herd boar's maternal background before plunking down hard cash for him, because it is those maternal genes that will be most directly expressed in the gilts he sires. Viewing the boar's dam and even granddam would be a valid part of any selection process, but on some farms this may not be possible, because the breeding herd is kept closed and isolated as a health care measure. In any event, when questioning the breeder do not overlook the maternal performance behind any boar you are considering.

As a prospective boar buyer you need to make inquiries as to the size of the boar's litter at farrowing and at weaning, how well the sow milked and mothered (this is often demonstrated by whole-litter and individual pig weights at 21 days of age and again at weaning), and how many litters that female has produced during her life in the herd.

Sizing Up Genetics

When I go out to buy a new herd boar, I look for the biggest pig from a big litter and one that has at least two or three keeper-quality male siblings among his littermates. I consider this an indication of both the depth of the genetics behind the mating and the dam's ability to raise a quality product in good volume. Producers working with smaller numbers of animals must emphasize this level of production and consistency in order to stay competitive.

Soundness and Appearance

Soundness is an absolute must in boar selection, as his success as a breeder depends greatly upon his agility and athleticism. Hogs with hind legs tucked too far under the body, knocked knees, sickle hocks, or other foot- or leg-structure problems should be weeded out early in the selection process. A good boar walks out on a lot of bone and will remind you of a champion boxer striding confidently to the ring.

A bit harder to pin down on paper, but also important in boar selection, are ruggedness and masculine character. Boars, especially purebred pigs, change greatly as they mature and develop — there is often even a time or two in their developmental stages when they may actually appear to be falling apart. Purebred hogs have long been bred to grow and develop a certain way

This is a fine young Hampshire boar. He has a masculine head, which is a desirable trait.

Hampshire Breed Association

and, as a result, they have their gawky moments until they emerge as a "finished product." It is a widely held belief that the two best times to select a boar are on the day he's born, and when he reaches 300 pounds and begins taking on his mature characteristics.

You should certainly select for large, fast-growing boars, but there also are a number of seemingly small guideposts that veteran producers use to select boars with lots of bred-in growth potential. Among them are width between the eyes and foot or jaw size. Many believe that the young animal will grow to match or "reach" those features. I opt for the width-between-the-eyes measurement, because I believe it corresponds to body width and good internal dimension.

Not only do male hormones influence growth, but they also go on to produce a number of secondary sexual characteristics. Among those are a larger, coarser, and more masculine-appearing head; more guttural vocalizations; and a chopping activity with the mouth that sometimes produces large amounts of a white, foamy substance. The foam appears to be a natural sexual attractant to sows and a young boar that doesn't do a lot of chopping and rattling gates in displays of sexual aggression is probably going to be a problem breeder.

It sounds almost too simple to need to be written down, but a boar should look like a boar.

Mark Hess/American Livestock Breeds Conservancy

A young Hereford boar in the breeding pen. Note excellent width between the eyes that suggests dimension of the respiratory tract through the body. Also obvious are a good depth of side and a well-defined ham.

Take Teat Count on Boars, Too

Checking the underline and making a teat count is important in boars, too. Teats are an important trait to emphasize in the selection of a breeding boar, because his stamp will be borne far more by his daughters than his sons. A minimum of 12 evenly spaced, large and well-formed nipples should be found along his underline. Teat placement on boars is also a good visual indication of carcass length; I always try to select boars with at least three teats ahead of the penile sheath on each side of the underline.

Reproductive System

Unlike the case of gilts, most of a young boar's reproductive system is open for visual inspection. The testicles should be large, well formed, and carried in the scrotum in a slightly asymmetrical position; that is, one testicle should be positioned a bit higher in the scrotum than the other. Avoid a boar with testicles that are noticeably uneven in size. There is some statistical evidence that the larger the testes, the sooner the male will enter puberty, and the more fertile he will be. The penile sheath should be trim in appearance and fit closely to the underline. In a too-pendulous sheath, fluids can collect

Chester White Breed Association

This is a big-framed Chester White boar showing excellent frame and muscling while maintaining good breed character. He's the kind of boar that's the small hog raiser's best friend.

and create a pocket ripe for infection, which might damage sperm quality. Injuries to the penis and bleeding from the sheath or penis are both solid reasons for rejecting a boar.

Tips for Buying Boars

Whether you plan to buy a boar at a breeder's sale or elsewhere, here are a few steps to follow to assure a successful purchase:

◆ If you're going to a sale, request a catalog and study it in some detail. Use it to compare the performances of the various bloodlines represented, and to determine how consistently a line you're considering performs. It is a tool for the comparison analysis of available data and not simply a wish book.

◆ Call the producer ahead of time with any questions.

◆ Arrive early on the day of the sale to view in detail animals that you have marked in the catalog for consideration. This also gives you time for a short visit with the producer about the offering.

◆ Take advantage of the breed association and auction company staff on hand. They are there to help you make evaluations.

◆ Remember that the order of the sale is generally based on merit, with the animals determined to be best selling first to help set price trends and the tone for the whole sale. Still, do not be surprised to see a littermate to the sale starter enter the ring much later and sell for a great deal less than his or her sibling.

◆ Have a walk around the whole barn and even a turn inside the sale ring. Sawdust or sand in the sale ring generally means hogs with a set of feet and legs to show off. A deep bed of straw in the sale ring can mean just the opposite. Also, look up all of the siblings being offered to the hog or hogs you are considering. If there aren't any or they're way down in the offering, let this temper your thoughts, because a really good line should produce more than one quality animal.

◆ Go to the sale with a ceiling price for your bidding firmly in mind and stick to it.

Boar Warranties

There are a few terms or conditions to boar guarantees, and they form a pretty good set of guidelines about the care and management of the young animal. Warranties will vary somewhat, a boar is generally warranteed under the following terms:

1. The young boar must be used in a hand-breeding situation for the first few matings. He must also be brought into contact with one female at a time, preferably a young sow or gilt close to him in size, and in good, standing heat. Early trauma such as fighting with a larger female not in estrus can damage or destroy a young boar's breeding career. Many breeders reserve the right to test the breeding abilities of a contested boar, however.

2. The warranty is invalid if the boar is simply dumped into a pen mating situation. We once sold a pair of five-month-old boars that looked big enough to breed, but really needed at least another 45 days of seasoning. When we delivered them, the buyer had us unload them right into a pen of gilts of several different ages. I'm afraid that buyer wasn't very happy when I told him those boars were strictly as-is-where-is, and that he was subjecting them to use and risks far out of reason.

3. The warranty is invalid if any rings are placed in the young boar's nose. A boar often uses his nose to position the female for mounting and service; ringing can create boars that are shy breeders.

4. Many breeders require notice as soon as possible that problems are emerging with boars they've sold. Some limit the warranty period to 30 to 90 days following the date of sale. We have had less than half a dozen complaints in our nearly 30 years in the boar business, and two of the most vociferous were actually due to producer neglect or slight. The most vocal was not about a boar's failure to perform, but uneven litter size at farrowing. That was quite clearly a management problem, since uneven litter size is generally a nutritional problem, environment related, or the result of a health problem that develops following the relocation of the boar. For example, even low-grade fevers at the time of breeding can adversely affect sperm counts in boars, and embryo survival in bred females.

5. The replacement clause in such a guarantee normally provides for the replacement of the boar with another of equal value, or a discount on the purchase of another boar.

A thick Landrace boar in motion. Note extension of front leg or "reach."

Final Decisions

There is one final question regarding breeding stock selection. What is a good one worth?

One of the founders of the legendary Wye Plantation herd of Black Angus cattle once stated that you should expect to spend at least the monetary value of your five best females to acquire a male that will truly advance the depth and quality of your herd or flock. Such a high expenditure may not always be necessary to maintain a high-quality swine herd, but it does point up the value that you must posit for additions to your breeding herd if quality is to be maintained. The time of widespread four-figure prices for swine breeding stock now seems to be well past. A true farmer/breeder now buys a whole lot of boar for $300 to $600.

Gilts are generally priced via a formula tied to current butcher hog prices. Typically, this will have a 250- to 300-pound gilt selling for market top at a specific market point (often one of the Iowa markets) plus $25 to $100 per head. Bred gilts seem to fall fairly consistently within a $350- to $750-per-head range, with the top end for pedigreed females carrying pedigreed litters. The Missouri Pork Producers group regularly awards $1,000 to a number of 4-H and FFA youngsters confident that they can each purchase two bred, pedigreed gilts at the state's annual spring-bred gilt sale.

Gilts bought in groups may be discounted a bit from the prices for gilts bought one or two at a time. Gate-cut gilts from finishing pens can sometimes be bought for a slight increase over what they would bring as butchers. And some good gilts can be bought at the various weanling pig sales.

Buy from Fellow Farmers

A bit of an aside here, but I would like to make a pitch for buying pure-bred seedstock, and buying it from your fellow family farmers. The so-called seedstock companies spend a lot of money on advertising and provide a few ruffles and flourishes like 800 phone numbers and delivery services, but they, too, tap into the swine gene pool that is held and maintained by America's family farmers. A great many of the lines these companies offer are crosses and breed composites, and it is the companies that reap the greatest share of the hybrid vigor benefits.

EIGHT

MANAGING YOUR BREEDING HERD

ONCE YOUR ANIMALS ARE SELECTED AND BOUGHT you cannot just take them home, drop them in a pen, and start counting off the days until the pigs arrive.

New breeding stock should be isolated for at least 30 days and, during that time, closely observed for any potential health problems. They need to be penned well apart from any other hogs on the farm and tended last each time you do chores, to prevent you from tracking any potentially harmful organisms from their pen to other animals.

It's not practical for small farmers to maintain a completely closed herd, but we keep our breeding stock on the opposite side of the farm from the animals for sale. Visitors generally don't see the breeding herd, and during farrowing season we pretty much just stay home, because the health risks are greatest for little pigs. Chore clothes and footwear are not worn off the farm. We do not visit other hog herds, nor travel to sales or other events where farmers are apt to gather, when we have farrowing sows and very young pigs on hand. And I very much discourage visitors in and around farrowing quarters and young pigs. More than one feed salesman has been run off a farm for entering a hog building unbidden or even driving onto the wrong part of a farm.

Bringing in the Boar

Especially critical is how a young boar is brought into the herd, because even a slight injury or low-grade fever can knock him out of service for a month or more. Such a time lapse can badly skew breeding schedules and disrupt both farrowing plans and budgets.

There are a number of simple steps, however, that will help you ease a young boar into the breeding herd:

1. *Boost immunity.* After the quarantine period is safely past, many hog farmers transfer a few scoops of manure and old bedding from the sows' pen to the boar's pen, and vice versa. Every farm has its own population mix of good "bugs" and bad, and this early exposure can prevent later health problems, especially fevers, which can cause temporary sterility in boars (or the failure of embryos to implant in a sow's uterus).

2. *Allow fence line contact.* The next step for many is fence line contact between the sows and boar. Pen the boar adjacent to the females so that they can sniff and have nose-to-nose contact with each other through the fence wire. They become more comfortable with each other this way, and the chances of fighting when the new boar enters the breeding pen are greatly reduced. Fence line exposure can also cause a pen of females to begin cycling. These must be very stoutly made pens: 54-inch-high cattle panels set with posts on 8-foot centers are probably the best choice for their construction.

3. *One-on-one introduction during estrus.* The first few gilts or young sows in estrus should be exposed to the new boar one at a time, and only when they are in standing heat.

4. *Observe.* You should be on hand to observe those first few services to detect any problems. Things to note especially include high riding, which can cause irritation and injury to the boar's penis. Look also for blood and ejaculate on the sow's back, on the penis when extended, or on the sheath. Blood kills semen. High riding generally is grounds for immediate culling, and an injured penis can be slow to heal. But some high riders do learn better technique through trial and error during hand mating.

Fence line contact between the sows and boar is one way to ensure that the animals become more comfortable with each other, and less likely to fight when the new boar enters the breeding pen.

How Many Boars?

The rule of thumb is to maintain one boar for every 10 sows on the farm. And even a young boar should be able to breed and settle half a dozen sows or gilts on a single cycle. Double-breeding the female at 12-hour intervals during her estrus cycle, or breeding her to two different boars, will also increase successful breeding percentages.

You've probably heard about artificial insemination (AI) in the swine industry. Personally, I do not favor AI. Semen has to be used fresh and the cost for purchase and transportation can be expensive. Also, the semen trade is in the hands of a few corporations now, and in my view, AI hurts boar sales for small producers.

Bringing in Gilts

Trying to bring one or two gilts into an existing group of older sows can be very trying; in fact, many producers choose to replace whole farrowing groups rather than attempt piecemeal changes. If timed to coincide with periods of higher sow prices, the salvage value of a heavy cull sow is often enough to buy her gilt replacement. If you retain and raise your own gilt replacements, keep and breed at least 10 percent more females than you will actually need, to be sure an adequate-size group will be in sync to farrow together.

New animals being introduced to a group should be closely watched for at least several hours. If a young female is too timid, she may never meld into a group, and her entire lifetime performance will suffer as a result. If you find yourself with such an animal, pen her separately, put some weight on her, and send her to town.

Be vigilant about watching new gilts at feeding time for several days, and consider using feeding stalls to confine animals individually. These stalls generally are 24 inches wide by 72 inches deep. They are made of native hardwood or treated lumber, and are normally erected with 2-inch flooring on runners for easy portability. They can be made with or without a plank floor, but hogs usually tread the ground muddy beneath crates without floors.

The safest and most useful stalls can be sealed with a gate after the hogs enter them. Thus each female can eat safely, take all of the time she needs to eat, and receive exactly the amount of feed she needs; also, younger and smaller females aren't dominated and harmed by the more aggressive "boss" sow(s) in a pen. The stalls are also useful for containing females to receive individual health treatments.

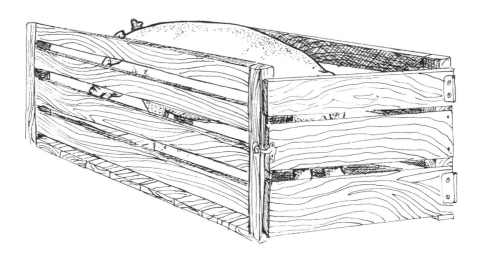

The safest and most useful feeding stalls can be sealed with a gate after the hogs enter.

Integrating Gilts into a Small Herd

Gilts are younger, generally smaller, and more docile than sows, and if they are simply dumped into an established pecking order there is a very real risk of serious injury to the young animals. To fill in one or two openings in a very small group of females, there are a few measures to try, although none are a guarantee of success. These include:

- Moving the sows to a pen where the gilts are already established
- Spraying all of the animals with a strong-smelling liquid such as kerosene, so that they all smell the same to each other
- Penning the animals in a very large enclosure with plenty of running room and extra sleeping sheds (this is our preferred option)
- Using feeding stalls

Breeding Gilts

Gilts generally enter puberty at around eight months of age (six to eight for boars) and cycle every 21 days thereafter until bred. To get gilts ready to breed, many use a practice called "flushing." For 7 to 14 days prior to entering the breeding pen, the gilts are placed on full feed — roughly 3 percent of their body weight daily. For gilts that have been limit fed to maintain condition, or sows that have been dramatically nursed down, this gives their systems a sort of kick start.

As I mentioned above, exposure to the boar can often start a set of females to cycling. Another old trick to get gilts cycling is to take them for a short haul in the pickup or trailer. Many farmers will pick a nice morning, load up the gilts, go to town for a cup of coffee, take the long way home, and then unload them into the breeding pen. This break in the usual routine seems to trigger responses in the gilts.

Other than watching for the boar's response, one way to tell if a gilt is in estrus is the back-pressure test. In the presence of the boar, put pressure on the gilt's back. If she's in estrus, she'll "stand." Her ears will come up and her tail will twitch.

Just ahead of breeding is also a good time to treat the animals for internal and external parasites. (Check out chapter 9 for more information on this.)

A few hog producers "hand breed," which means placing each female in standing heat together with the male, witnessing the mating, and returning the female to a separate pen. They will then breed the females again in 12 or 24 hours. For this effort, you'll have an exact breeding date, but the process entails added labor. In addition, sows penned away from boars may be slower to come into heat and have less pronounced estrus cycles. Some sows may even experience a "silent heat," with no detectable signs of estrus.

Mating Success

How do you tell if mating was successful? Following a completed service, a "semen plug" may appear in the vulva. It is formed from ejaculate material and prevents semen from leaking out of the reproductive tract. There may also be a rather pronounced flour paste smell. Also look for dirt and hair disruption on the sow's back, which indicates that mounting occurred.

Gestation

A sow is pregnant for an average of 114 days — 3 months, 3 weeks, and 3 days, as the old hands like to relate it. During that period a gilt needs to gain about 125 pounds to ensure her continuing growth and good litter size at farrowing. A sow in good flesh needs to gain 50 pounds less during gestation. A nursed-down or very thin sow, or a second-litter gilt, may need to gain more than that additional 75 pounds. Added feed or increased fat content in the feed may be in order.

During the first two-thirds of gestation, embryo growth is quite modest, and most sows can get along quite well on a simple maintenance ration. This can often be met with just 4 pounds daily of a complete ration with a crude protein content of 14 to 15 percent. They should be provided with about ½ pound of crude protein daily. In very cold or raw weather, the females might need an extra pound of the complete ration or just corn to help maintain condition.

In the last third of the gestation period the fetal litter makes its greatest surge of growth. The demands on the sow's bodily reserves increase, and these demands will continue throughout the lactation period. During this stage of the gestation, daily feed levels may have to be increased by as much as 1 to 2 pounds. It is also the time to bring out the highest-quality feedstuffs.

Watering Breeding Stock

Most hogs today are watered with some sort of tank- or vat-type waterer that maintains drinking water before the hogs at all times. Kerosene burners or electric heaters are normally used to keep the waterers open in winter; painting them black will help the tanks draw a bit of solar heat. Offering your hogs all of the water they can drink in 20 minutes twice a day will also work if there is plenty of trough space and no animal is forced back.

Prefarrowing Cleanup

Two to four weeks ahead of farrowing, you should begin a cleanup process to ready the females. This begins with a thorough treatment inside and out for parasites. Again, check out chapter 9 for more details on worming.

The second stage of the cleanup moves to the farrowing quarters. During my Future Farmer of America days, it showed up on our grade card when a surprise visit by the ag teacher revealed dirty houses, crates, or stalls. Before a new set of sows went in, the farrowing quarters were to be cleaned and scraped down to bare wood or metal. Then they were saturated with a strong disinfecting spray solution such as a diluted iodine.

In an earlier day they would have been sprayed or scrubbed down with a strong lye solution, or even smoke-fumigated. Why anyone would ever want to eat off a hog house floor was never made clear to us, but the grade book made sure that we could.

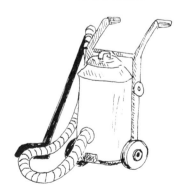

A steam cleaner is a good way to clean farrowing quarters.

Current thinking is that a simple cleaning and scraping were generally more than enough. Dumping farrowing house wastes into the gestation pens a month ahead of sow due dates, and getting the sows into the farrowing quarters a week ahead of their farrowing dates, will help each animal develop her own natural immunity to any harmful organisms. It is also far easier than trying to sterilize all of the porous surfaces and nooks and crannies in a farrowing unit. Modern steam cleaners and pressure washers make farrowing quarters cleanup even easier.

Farrowing

Get the sows into the farrowing quarters a few days early so they can settle in and develop a comfortable routine. Trying to move a sow too close to her farrowing time can leave her unsettled and greatly increase the number of pigs lost due to overlay or other fretful behavior.

The thought of taking a sow or sows through the farrowing process has probably put off more would-be pork producers than any other aspect of pork production. Actually, it is one of the simplest, most basic and natural of occurrences, and generally passes with no need for excessive human intervention.

If you have selected large, well-grown females, fed them well, maintained their health, and matched the right type of boar to them, there is really very little room for problems. In 30 years at this hog game, I can recall just a handful of rough farrowings, and the vet was able to talk me through a couple of them over the telephone.

Signs of Labor

As the time nears for the farrowing, the sow will give a number of clues to the imminent birth of her litter. She:

◆ Will become restless
◆ May display nest-building behavior
◆ Will have milk let down
◆ May have a slight vaginal discharge, but not always
◆ May circle about her pen and sniff behind her often

When you see these signs, the sow is about to begin heavy labor. Once farrowing starts, the pigs should be born at roughly 20-minute intervals. The smallest pigs will be born first and last in the litter. An old rule of thumb holds that small litters will come late and be mostly boar pigs; large litters will arrive early and be mostly gilts.

Trouble Indicators

The sow's behavior and actions serve as good indications of something that's going wrong. Three or four hours of nonproductive labor or obvious signs of great pain or stress on the sow's part let you know that intervention is needed. In most instances a sow can safely deliver a breach or dead pig on her own, but this will slow down the birth process. The same is true of a very large pig, and if the pigs behind it are caught too long in the birth canal they can, quite literally, drown.

Other signs of a problem delivery include a sow that gets up and down often, a foul-smelling or bloody discharge from the vagina, and a pig's limb or head (especially with an extended or blue-appearing tongue) that has only partially emerged.

When problems become evident, the quicker a veterinarian is called in or other intervention made, the better. More pigs may be saved, the sow will fare better, and your costs should be less in the long run.

I have, however, entered a sow myself to assist with a birth or just to better determine how things are progressing. When it is necessary to enter the birth canal there are a number of basic steps to follow:

- Clip all of the fingernails on that hand short.
- Thoroughly clean your hands and arms to well past the elbows.
- Lubricate your hand and arm (mild liquid soap will do in a pinch).
- Shape fingers and thumb into a sort of bird's beak.
- Enter through the vulva, and proceed slowly until any obstruction is encountered.

There is not a lot of room within the birth canal, so you have to proceed quite slowly. When you encounter a pig, you must determine its position to assist in its birth. Grasping a pig in the birth canal can be difficult; there are a number of stainless surgical snares that can be sterilized and used for pulling pigs. Again, in a pinch, you can even boil a bit of baling twine and use it to form a simple obstetrical snare. If possible, try to grab the head; if it's a breech birth, draw the pig out by the hind legs.

Prolonged labor can tire a sow and greatly slow her contractions. You need to get the obstructing pig out quickly, but not in a way that will injure the pig or sow. Try to time your slow and steady pulling to coincide with the sow's contractions. Pull slowly and steadily and with a downward cant.

Oxytocin is a drug that can stimulate uterine contractions and milk let-down. It may also be one of the most-abused drugs used in modern livestock production. Some use it just to speed up the birth process; others, without checking the birth canal for the cause of the slowdown. With a very large or dead pig blocking the birth canal, the drug may do little more than wear the sow down further. Also, it is easy to give an overdose of this drug. Our vet recommends just one or two injections of no more than 1 to 2 cc's each. It is a prescription drug.

Often, getting one pig moving will again get the birth process going apace. The pigs should then begin arriving at regular intervals and, after that, the placenta be expelled. If the sow fails to "clean" — that's a hog-producer's term for shedding all the placenta — a severe infection can develop.

Anytime the sow's birth canal is entered she should receive a substantial injection of a long-lasting antibiotic such as LA-200. I give 10 cc's on each side of the neck.

There is a very real question as to how much involvement you should have with a normally farrowing sow. A healthy female with good body capacity should be able to lie down and farrow at least eight live pigs, if a gilt, and at least nine if a sow. You can be a bit more forgiving with a gilt or a purebred litter, but most of the females that miss this mark should earn a one-way trip to town. I now bed our piggy females down following evening chores and generally don't go back out until morning. In short, with a normally farrowing gilt or sow, your involvement will be minimal.

The farrowing process is one of the simplest, most basic, and most natural of occurrences, and there is seldom need for human intervention.

U.S. Productivity Averages

In 1995, the USDA Animal and Plant Health Inspection Service conducted a major study on swine. It included figures for farrowing and weaning productivity in the United States representing 95 percent of the U.S. swine herd. Some of those findings follow:

Average number born/litter	10.22
Average number stillbirths or mummies	0.76
Average number born alive/litter	9.50
Average number weaned/litter	8.61

The following figures show the three leading causes of preweaning deaths, as identified by producers:

Laid on by sow	47.7%
Scours	16.1%
Starvation	15.4%

An extension swine specialist involved with the report says that the preweaning death rate of baby pigs, which is less than 10 percent, is encouraging. It was once estimated to be as high as 20 percent. The report also underscores the importance of providing housing that will protect little pigs from overlay.

The Pig at Birth

Most pigs are born with their eyes open, good reflexes, and a strong will to nurse and live. Many hog farmers used to and still do stay by the sow's side to dry off the newborn pigs and place them on a teat. I have some most pleasant memories of long evenings in the farrowing house with Dad, the air rich with the smells of straw and his pipe tobacco, and discussing the merits of each pig as it was born.

A watchful eye can save the occasional pig. Some are born in a saclike membrane; unless they are freed from it, they will suffocate. Sometimes encountered is the odd female that becomes tense at farrowing and may savage her newborn pigs. Studies indicate that savaging of piglets may be more likely to occur in first litters, and that the victim is more likely to be the firstborn. It also has been shown that sows that demonstrate this behavior were more likely to have been bred at low weight and in a depleted condition. Nevertheless, such females should be culled quickly and their offspring never allowed to enter the breeding herd, because they may carry on this nervous behavior and newborn savaging. Newborn pigs can be withheld from their dams for up to four hours following farrowing for their protection or to be taken somewhere for warming.

American Livestock Breeds Conservancy

Most pigs are born with a strong will to nurse and live.

Providing Breathing Help to Newborns

Pigs that needed extra time to be born or that were born in a membrane may initially need help breathing. There is a simple procedure to help them gain that all-important first breath.

- Clear the pig's mouth and nose of any fluids or membrane.
- To remove fluids from the pig's mouth and lungs, suspend it head down and slowly swing it back and forth between your legs.
- Again clear the mouth and nose.
- Dry the pig with clean rags.
- Set it down to nurse that all-important colostrum.
- Monitor the pig's condition closely for the first few hours following its birth.

Caring for the Young Pig

In the first three weeks of its life, a pig is a very different creature from the animal it will become. It is much more temperature sensitive and almost totally dependent upon an all-liquid diet. Pigs with a birthweight of 3 or more pounds also seem to have a far greater chance of survival than pigs with lighter birthweights, especially weights of less than 2 pounds.

Little pigs will nurse about 16 times a day for several minutes at a time. Their very nursing activities seem to stimulate milk release, and a contented sow will make soft, guttural noises while the pigs nurse.

Young pigs often develop diarrhea, which also is called scours, and can soon dehydrate and all too quickly die. At about three weeks following farrowing, a sow's milk flow peaks. It is about the same time that she would be entering her estrus cycle were she not nursing. Many believe that this may even key a chemical change in the composition of her milk. Such subtle changes can trigger scouring or increase stress in pigs not being well and carefully tended.

Too Hot or Too Cold?

Temperature extremes can be deadly to young pigs, so watch for signs that they are chilled or overheated. Young pigs that are piled up and have their haircoats on end are chilled and huddling together for warmth (pigs at right). They are obviously under stress. If they are spread out and panting, they are too hot (pig at left).

A swine disease called TGE (transmissible gastroenteritis) is especially insidious as the pigs may both scour and vomit, and the death rate in very young pigs can approach 100 percent. The fecal odor with a TGE outbreak is quite distinct.

More on scouring and TGE appears in the next chapter, but I'll say here that the best course of action to take with TGE is to counter the symptoms and build natural immunity into the next set of sows due to farrow. If they are 30 days or more away from farrowing they can be exposed to contaminated bedding and fed the contents of the intestines of pigs stricken with the disease.

The real trick is to keep such health problems away from the farrowing units and the breeding herd in the first place.

Colostrum

Colostrum — the all-important first milk — imparts natural antibodies that the young pig must have to assure its very survival. In fact, it is virtually impossible for a pig to survive if it does not receive any colostrum.

Colostrum is secreted in the first 12 to 24 hours following farrowing. A pig needs to nurse within four hours of birth. A pig that isn't nursing will weaken, grow cold, and may move away from the sow to a distant point in the crate or stall. A pig that survives the first day without receiving colostrum will fail to thrive and may die many days later.

Goat colostrum is probably the best substitute for a sow's, but cow colostrum will work. Sow's milk may be the richest given by any domestic animal. Replacement colostrum can be frozen in an ice cube tray; pop the frozen cubes into a plastic bag, and store them in the deep freeze for future use. Do not thaw the frozen colostrum in a microwave oven, however, or heat it with overly hot water; both can destroy vital antibodies in the colostrum, making it virtually useless to the piglets. Thaw it in a double boiler or at room temperature.

Initial Management Tasks

Within 12 to 24 hours of farrowing several management tasks must be addressed. These are listed below. Not every producer uses every one of these practices, but you'll probably want to employ at least some of them.

Grouping several treatments at the same time when the pigs are quite young minimizes stress. For instance, many hog farmers clip the wolf teeth, treat the navel, notch the ears, and dock the tails during a single session.

Wolf teeth clipping. There are eight wolf or needle teeth in the pig: two on each side of the upper jaw, and two on each side of the lower. They should be clipped off at the jaw line. For this task, many hog producers use small clippers or sidecutters.

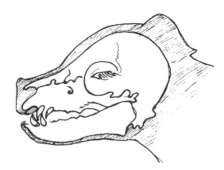

The wolf teeth.

If they're not removed, wolf teeth can cause such discomfort to the sow's udder that the sow may not let the pigs nurse. Wolf teeth also can be used by the little pigs to cut each other in struggles over space at the udder. Pigs with small, crusty-looking scabbing on the snout and face are victims of needle tooth lacerations.

Navel treatment. Treat the navel with an astringent, such as an iodine solution, wound spray, or Blue Lotion. Along with cuts and scrapes at the knee joints, the navel is one of the primary routes infection can take into the very young pig.

Ear notching. Young pigs often have their ears notched, clipped, or punched as a means of individual identification. This is useful for age verification, as well as positively linking the pig to its sire and dam. Many producers also put a simple notch or punch a small hole in one ear of gilts from big litters that they wish to denote for future consideration as female herd replacements. If done in the first 12 hours following farrowing, ear notching is quite simple and causes the little pigs very little stress and pain.

Right ear is notched to identify litter; left ear shows pig's individual number in the litter.

The ear-notching method shown here is the one most commonly used, and is required by most of the major swine registries for the issuing of pedigree documents. Some states may require this type of individual identification for certain disease-control measures and the issuance of transport documents.

Tail docking is now a somewhat controversial procedure, because animal activists charge that it causes pain. But in pigs, as in newborn dogs, the tail right after birth is simply soft gristle, and docking it appears to cause minimal, if any, discomfort. In fact, there is virtually no bleeding.

Feeder pigs are often sharply discounted at auctions if they are not tail-docked. Pigs crowding in to a self-feeder or in fairly close confinement may engage in tail biting to move other pigs out of their way. Such activity can create sites for infection, and in extreme instances has even led to cannibalism.

To avoid this, then, use sidecutter pliers or nippers to remove the last half or two-thirds of a pig's tail at the same time the pig is ear notched and has its needle teeth clipped. Clipping the tail any shorter than this may lead to problems with rectal prolapse, in which the rectum actually slips out of its proper place.

Many hog farmers will not dock gilts that are under consideration as female herd replacements. And we have had a problem selling boars that have lost a portion of their tail for whatever reason, because it is considered cosmetically undesirable by some.

Don't clip the tail too short; it could lead to problems.

Castration

Boars not intended for stud service should be castrated because the hormones that emerge when they reach maturity could taint their meat. You'll want to castrate soon after birth, however, not as they're reaching maturity, because that's when the animals are easiest to handle and will heal the quickest.

Castration may be an even more controversial procedure than tail docking. Still, in this country the meat from even the youngest uncastrated boars is sharply discounted in price. In Europe a lot of lightweight hogs are used for fresh and processed pork products, many of them intact males of five months of age or less. They are slaughtered before many of the more troublesome secondary male sexual characteristics can develop. These are triggered by those male hormones mentioned above and can range from coarse hair and the development of tusks to the so-called boar taint that the packing industry so fears. In the meat, it manifests itself as both a strong cooking odor and an unpleasant taste.

Over the years, we have eaten pork from a number of young boars. As long as they were not in service, and were processed at a reasonable slaughter weight and age, we had no problems with the meat. Our local processor, however, will not render their fat for lard lest there be some slight trace of taint that might affect a whole kettle of lard.

The earlier a young male is castrated, the better, and the less stressful the procedure will be. Although it is still sometimes tried, castrating a boar that has been in service is pointless. Because I produce boars for sale as breeding stock, I have waited until as late as 10 weeks of age before castrating the last of a group of young males.

Initial Injections

I give pigs a ½ cc injection of a long-lasting antibiotic, to counter any potential problems with navel ill or other early infections. The injection site is one of the long muscles in the side of the neck. (More information on how to give an injection appears in chapter 9.)

In the other side of the neck can go a 1 to 2 cc supplemental iron injection. A sow's milk is richer than that of any other common farm animal and would be the perfect food were it not lacking in iron content. Without that boost of iron, the pigs may soon develop a problem with iron deficiency anemia.

I and many other hog producers use a 1 cc injection initially, then give a second when the pigs are 10 to 14 days of age. And another iron injection may be in order at weaning or shortly before.

There are alternatives to injectable iron. Some iron products can be given orally via a pump. And there is a group of products commonly called "rootin' iron" because the mineral is in dry form, mixed with a peat moss carrier that pigs actually root through. I prefer the injectables, then the oral products, because they ensure that each pig gets an exact amount of the needed iron. For the "needle shy" the oral pump bottle may be appropriate.

Castration procedure. Castration can be done through the scrotum. It works with pigs of all ages — although with animals of any size, it requires two people. Place the larger pig on its side on a table or straw bale; one person, serving as a holder, applies pressure with a knee or arm just ahead of the ham to hold the hindquarters very steady.

The necessary incisions tend to be rather small and exacting. Make a small cut over each testis and through the scrotum; draw the testis free and sever it along with a bit of its supporting cord; then treat the surgical wound with an antiseptic astringent or wound protector.

To make the incision, you can use a single-edged razor blade, pocketknife with spay blade, scalpel, or one of the new disposable, hooked-blade castration knives (about $1 each and adequate for 10–12 castrations if properly used). An aerosol spray antiseptic is very useful for treating the surgical wounds, because it penetrates them quite deeply.

There is another method of castration, and with it one person can actually castrate a pig of up to 60 pounds or so. It is sometimes called the "show barrow cut," because it leaves no visible scarring. The method calls for a pig holder, which is a device that slips onto a gate top or stall partition.

Slip the pig's hind legs through members on the holder, which suspends the animal head down and with its belly facing out. The testes move downward between the animal's hind legs. Place a knee between the pig's front legs to hold it steady, and grasp one testis at a time between thumb and forefinger to bring it taut against the skin (this causes no pain).

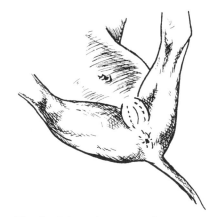

Make a short incision with a hooked castration blade just above the testis. The pressure you are applying beneath it with the thumb and forefinger of your other hand should cause the testis to pop through the incision. Draw it out, along with a goodly portion of the cord-type system that supports it. Then sever the cord to free the testis. You can treat the wounds at this point with a good antiseptic spray. Many find the testes to be quite tasty fare.

The incisions for castration are small and exacting. They are made over each testis and through the scrotum. It would be wise to enlist the help of your veterinarian or an experienced hog producer to teach you the castration procedure.

Postcastration Care

Freshly castrated pigs should be returned to a dry, freshly bedded stall or pen. There should be no mud for them to lie in, and water should be offered in pans or troughs in which the pigs cannot lie. This will keep water and foreign substances from entering the surgical wounds.

The pigs should be watched closely for several days for signs of infection, such as excessive scrotal swelling or fever. I have never had a problem following castration; the wounds heal and close quickly.

Sometimes, however, a hernia results after castration. One of the animal's scrotum will drop lower than the other, appear larger, or otherwise be asymmetrical. In this case the intestines must be returned to their proper place and the site sutured, so you may well need the help of your veterinarian.

Postfarrowing Sow Care

On the day of farrowing many hog farmers totally withhold feed from the sow, the idea being to circumvent gastric upset or constipation, which might disrupt milk flow.

For a few days before and after farrowing some also add bran to farrowing rations, as a further preventive measure or topdress the ration with oat bran, or with beet pulp, which comes in a meal-like form. Some of our Amish friends offer warm water with Epsom salts in it for the first drink following farrowing.

It is important to note when a sow cleans, rises, and begins normal bowel function following farrowing. Getting her up is critical. A few sows just flat fail to rise for no really good reason — perhaps a fear of a wet, slippery floor — and this may be one of the few times when the use of an electric prod is justified. You must get her up following farrowing. If you don't, she won't eat, and she'll eventually be lost to a combination of stress, shock, pneumonia, and pure cussedness. I want my sows up as quickly as possible following farrowing — within one to two hours.

Virtually from the moment of farrowing you must monitor the sow for signs of the MMA complex: mastitis, metritis, and agalactia. A uterine infection may cause a fever to set in, the udder to redden and harden, and the milk flow to decline or even stop. Be on the watch for vaginal discharge, reddening and stiffening of the udder, and excessive squealing or gauntness in the pigs, which indicates the sow is nursing poorly and the pigs aren't getting enough to eat. The condition may actually become so painful that the sow will lie on her belly to keep the pigs from nursing, causing her further discomfort.

Normal treatment procedures for MMA are antibiotic injections, uterine infusions, and oxytocin, sometimes called "pop," to stimulate milk letdown. Just remember not to abuse the latter.

Feeding the Nursing Sow

In most instances, we feed our breeding animals just once a day, in the late afternoon. In summer, the cooling temperatures at that time seem to stimulate appetite and in winter the sows go into the long, cold night with full stomachs.

However, the demands of nursing a litter require increased feed levels for the sow. About 12 to 24 hours following farrowing, many producers offer the sow about 2 pounds of her normal ration; then they gradually build the amount fed until the sow is on full feed (about 3 percent of her liveweight), over the course of a week or so. One old rule of thumb holds that within a week or so of farrowing the sow should be receiving 3 to 4 pounds of feed a day for her own needs, and an additional 1 pound for each pig she is nursing.

Some nursing sows may need as much as 16 to 20 pounds of feed per day in the latter stages of the lactation period. Certainly feed to appetite. The old rule of thumb noted above goes on to advise 4 to 6 pounds for the sow during late lactation, and 1 for each nursing pig. Generally speaking, the more you can get into the sow at this time, the better.

Many hog producers still divide up the sow feed and give it to the breeding animals twice a day, morning and evening. To get enough feed into lactating sows with large litters in warm weather, three-times-a-day feeding may be in order, especially when it's very hot — and then even three feedings may not be enough to get a sow to consume all she really needs. The heat suppresses metabolic functions as well as appetite. You can also add fat to the ration at the rate of 3 to 5 percent of the total amount fed, but you must do so gradually to prevent gastric upset. Some producers even top off the sow's daily ration with 1 to 2 pounds of milk and nutrient-rich pig starter if she is nursing an especially large litter. (Pig starter is discussed below.)

The nursing sow should also receive free-choice drinking water. Not only are demands being made on her by her nursing young, but she must also maintain body condition adequate to allow for a rapid breedback following weaning. Females that are neither nursing nor pregnant are costing their owner money. A hollow sow is a mighty expensive creature to own.

> ### Milk Production
>
> A variety of factors, such as general body condition, can affect milk production. But the most important factor is good nutrition. Make sure your sows are on a proper diet to help ensure that they produce enough milk.

Food Starters for Nursing Pigs

The next major decision in nursing management is how and when to provide pig starter — a ready-made, store-bought product — or even if it is really necessary to provide such complex feed if the sow is nursing well. Pig starter is perhaps the most complexly formulated of all swine feeds: It is high in both fat and protein content, is richly flavored, and has a substantial level of milk product. It is also the most costly, and rations for very young pigs, or prestarters, are often still more complex and costly. There is even a category of liquid pig starters and ration boosters designed to closely approximate sow's milk.

Trying to get the various starter feeds to the nursing pigs can also be a bit of a challenge. I have seen sows in crates literally turn back flips or stand on their heads to reach creep feed if they can. Those little metal and plastic crate pig feeders don't last very long when a sow gets to them.

I recall one noted midwestern Yorkshire swine breeder who pioneered the weaning of pigs at 35 days of age. It was his contention that with 35-day weaning, if a sow was worth her salt at all as a pig raiser there was really no need for pig starter to be offered to the nursing pigs. He called it little more than expensive pig bedding.

There are now multiple-stage pig starters intended to be fed to nursing pigs at various stages of their life, beginning just a few days after farrowing. Some are even meant to be fed at the rate of no more than a couple of pounds per head before changing to the next feed in the program. The greatest justification for such feeds is in early-weaning situations — weaning the pigs at 10 to 21 days of age.

With later weaning — six to eight weeks — a starter/grower feed can be useful in maintaining growth in the pigs, because the sow's milk production begins to decline three to five weeks into the lactation period. These simpler feeds still pack a bit of milk and flavoring, but typically have crude protein levels in the 16 to 18 percent range. Some of these early starters and prestarters crowd 30 percent crude protein and have such exotic ingredients as spray-dried blood product.

For later-weaned pigs, especially those in group situations in drylot or pasture, the walk-in creep feeder made of wood and sheet metal has been in use for decades. In theory, such feeders admit only the pigs, blocking out the sows that can consume many dollars' worth of the pricey feed in just a few minutes. In the real world, though, some old sows can play these feeders like a hustler plays a pinball machine. For best results, gate off such feeders when in use, to protect them from the sows.

On average a young pig will consume 40 to 50 pounds of starter/grower, along with mother's milk, in growing from birth to 40 to 45 pounds. By two weeks of age the pigs will also be robbing a bit from their mama's feedstuffs.

Sources of Stress

Throughout this book I emphasize the importance of minimizing stress on pigs and hogs, because stress lowers animals' resistance to disease. Sometimes, of course, it is not possible to completely avoid stress — such as the stress of weaning. Still, always do whatever you can to minimize it. And there are other types of stress that can be prevented entirely, such as overcrowding. Here's a list of the various stressors that can affect swine:

- Crowding
- Inadequate feeding or watering space
- Temperature changes
- Introducing new pigs
- Feed changes
- Weaning

Weaning

As noted, pig weaning now sometimes occurs when the pigs are as young as 10 to 14 days of age. Quite a departure from the traditional 56-day weaning age! Over the years I've also seen some pigs left on the sows until they were as much as 10 weeks — or 70 days — old.

I found that 56-day weaning took a greater toll on our purebred sows than I really liked to see, and also made it rather difficult to attain five litters farrowed within a 24-month period. Thus I tried 42-day weaning, then dropped down to weaning the pigs at 35 days of age. Weaning too early — under 21 days postfarrowing — has been found to cause problems for some sows with rebreeding. It seems to run almost totally counter to any natural cycles.

Our 35-day-old pigs are good size, have been venturing away from Mama and their houses for some time, are swiping regularly from the sow's feed, and are hardened off and acclimated to outside air temperatures. They can handle the less-complex, more moderately priced starter/growers, and have been raised by their dams long enough to give us a good indication of the sow's milking and mothering abilities.

Feeding the Newly Weaned Pig

Weaning can be one of the most stressful periods in a young pig's life. For at least one week before and after there should be no changes made in its diet. When changes in a pig's diet are made they should be done gradually, over a period of three days or so. In the first feeding offer a mixture of about one-third new ration and two-thirds current ration (which is the pig starter or a second-stage starter/grower feed). On the second day mix the new and old rations half and half. On the third day blend two-thirds new ration with one-third old. Pigs are, by the way, free fed: Offer as much as they want to eat.

Early growing rations generally have a crude protein content of 15 to 18 percent, and, if bought by the 50-pound bag, come in the form of small pellets or crumble. They are highly palatable and easily digestible. It is possible to formulate early grower and even starter rations on the farm, but some of their components have a quite short shelf life and are expensive to buy in small lots.

The cost and complexity of these rations is generally justified by the rapid and very efficient way in which young hogs grow. They can be the most positive investment you make in swine feedstuffs.

Maintaining the Newly Weaned Pig

Seas of ink have been expended on discussions of housing and maintaining the newly weaned pig. Very early-weaned pigs have to be moved into buildings capable of maintaining rigidly controlled environments. There they are kept in temperatures often exceeding 80°F, fed elaborate rations (sometimes even flavored with chocolate or strawberry), and held in single-litter pens, coops, or tublike structures.

Two hot buzzwords these days are *SEW* and *MEW*. SEW stands for "segregated early weaning" and is a measure taken to break swine disease cycles. The pigs are weaned when they are only about 10 pounds, which is quite young. They are moved to a nursery facility far removed from any other hogs — often a facility all alone on another farm several miles away from the rest of the hog herd. The idea is that distancing the young pigs from all other hogs keeps them from exposure to disease via these hogs or the environment that surrounds them.

MEW denotes "medicated early weaning." Also a practice intended to break disease cycles, it was developed in the United Kingdom and modified for use in this country. Basically, sows are immunized against several diseases before farrowing to enhance colostral immunity, and weaned pigs are moved to an isolated nursery unit, where they are separated from pigs of other ages. The separation by age and isolation appear to be more effective in controlling disease than medicating, according to Dr. Ross P. Cowart, author of *An Outline of Swine Diseases*.

Both SEW and MEW are quite costly practices to pursue. They are really outgrowths of the high-volume approach to swine production — the factory farming of pork, if you will. In such operations the swine-raising facilities are in virtually constant use, large numbers of animals are held in a very small area, stress loads on the animals are far greater, and individual animals don't receive adequate producer focus. The animals are relentlessly pushed to keep production on a factorylike schedule, and to keep the facilities full. All of this sets the stage for near-constant struggles with the "swine disease of the month."

On the small farm, however, where hogs are a part of a diversified enterprise mix and weaning ages run later, nursery accommodations can be far simpler. In fact, the least stressful way to wean pigs is to pull the sows and leave the pigs where they've always been. They are most at ease and comfortable in those surroundings. Note that your fencing must be quite pig tight to contain the pigs during the weaning period. They will walk the pen perimeters steadily looking for an opening that will lead them back to Mama. The sows should also be moved beyond scent, sight, and earshot of the pigs to further reduce stress on both groups.

Managing Pigs at Weaning

At five to eight weeks of age newly weaned pigs are not exactly hothouse orchids, but giving them a few breaks will do much to ensure that they get off to the best start possible. Some steps to help them through this potentially stressful time include:

1. Try to hold the newly weaned pigs in litter-size or penmate groups for at least the first few days following weaning.
2. Tie or solidly prop open lids on self-feeding and watering equipment, and closely watch over the pigs for several days following weaning. Newly weaned pigs may have trouble operating this equipment. It's normal for them to lose a bit of bloom or have a growth standstill for a short time, but watch closely for any that are becoming gaunt or losing tone or vitality.

   We once sold a set of just-weaned pigs to someone who hoped to save a few dollars by buying them young. He took them home to a large wooded lot with a creek running through it and a single, small walk-in feeder. The pigs found the water, but they had no experience with this type of feeder; nor were they familiar with the type of feed he was using (a complete meal rather than a pelleted grower). Several pigs actually died before the rest mastered the new feeder and ration. A call to our vet confirmed that this is an all-too-common occurrence.
3. Because pigs will drink when they won't eat, an inexpensive packet of vitamin/electrolyte powder mixed into their drinking water will help recently weaned animals maintain health and body condition. A box of flavored gelatin powder added to the drinking water will increase consumption levels.
4. For the first few days following weaning it may be best to offer feed-stuffs and drinking water in shallow pans or troughs. Feed presented this way can be changed often to keep it fresh and appealing.

Growout

With the stress and trials of weaning behind it, a pig begins its real work in life — making a hog of itself.

When pigs are about 10 weeks of age, hog producers begin blending litters. Feeding/finishing hogs are manageable in groups of up to 60 head or so. On most small farms the finishing groups typically contain the pigs from between 2 and 10 litters — 15 to 75 pigs per group.

Self-Feeder Space for Finishers

The rule of thumb is to provide one self-feeder lid or hole for every three to five finishing hogs in a pen or group. Most such feeders hold 40 to 80 bushels of feed (based on the 56-pounds-per-bushel weight of shelled corn) and have 10 to 12 feeder lids. Producers generally try to match even-size hogs, pen, and feeding and watering equipment to create a system that functions simply and requires only a modest amount of care and supervision.

Feeding to Finish

Hog finishing can, at first, seem to be little more than keeping the feeder and waterer full and watching the hogs grow, but there is more to it than that. Often much more. Here, I'll share my thoughts on hog feeding.

A finishing practice growing in acceptance is "split-sex feeding." This practice takes into account the different finishing traits of the sexes. Gilts grow a bit more slowly than barrows and tend to be leaner. They use protein to build muscle. Barrows grow faster, but tend to lay on more finish. As they grow, protein levels can be throttled back a bit in their rations.

Today's hogs are generally a lot better in type and lean content than those of even a decade ago and can do a first-rate job of utilizing high-performance rations. From about 60 pounds to market weight our Durocs and their crosses stay on a 15 percent grower — a pretty hot feed. It helps our animals to realize their optimum potential while not breaking the bank when we go to the elevator.

Never neglect drinking water quality and freshness. It is every bit as important as quality feedstuffs.

Farmers don't get up with the chickens any more, but watching finishing hogs come off their beds in the morning can be most helpful. This is a good time to spot respiratory ills, for instance, which often reveal themselves as coughing spells when the hogs first exert themselves. It is also a good time to spot injury and soundness problems. As hogs come up, their underlines and extremities are very visible. Those last pigs off their beds need an extra-close inspection to determine the cause of their sluggishness.

Be sure to provide the growing hogs with plenty of sleeping space. Eight square feet per head is a minimum, but 10 may be more in order. The animals need room to spread out in warm weather, and plenty of dry bedding in cold weather. In the latter case chilled hogs will pile up, and prolapses or even deaths by crushing and suffocation can result.

With small pigs, it may be necessary to block off a portion of the sleeping area until they grow larger. This is to prevent them from picking up the habit of dunging inside their sleeping quarters.

Seldom do all of the hogs in a group mature and reach handy market weight all at the same time; to be honest, I don't think I've ever seen that happen. You may need to top out a pen, which means removing hogs as they reach good market weight, two or three times over a period of weeks. This prevents your top-end hogs from becoming too heavy and receiving a price dock for carrying excess finish.

This may seem like a small thing, but we producers need to do a better job of how we refer to our growing hogs and finishing pen. A lot of time and dollars have been spent to make hogs leaner and more efficient, and a lot of this is belied every time a farmer is heard in a public place talking about "fat hogs," "fattening hogs," or the "fattening pen."

Handling Tailenders

In nearly every pen there will be a tailender or two. These are pigs that, due to injury, health problems, pure cussedness, or whatever, grow slowly or even stall out. A lot of them seem to hit the wall at about 170 to 180 pounds. You can drop them back to another group of younger finishers, but they seldom perform efficiently, and you can put yourself through a lot of stress and strain getting them blended into another set of hogs. These are good candidates for the home freezer or a trip to the local sale barn. They will sell for less, but it is often far better and simpler to just move them on and get your pen empty.

Nine

HOG HEALTH CARE

LET'S START BY SAYING THAT THE HOG is one of the most vigorous, prolific, and hardiest of all domestic animals. Day-to-day care of a small swine enterprise often takes just a couple of hours daily. It is largely a matter of keeping the hogs comfortable and content. When hogs are provided with good basic care, they will perform quite well with minimal special health measures.

Problems do arise, however, as with any species of animal. In this chapter you'll learn more not only about basic hog health care, which can prevent breakdowns, but also about how to spot the development of health problems.

A Health Care Philosophy

As I mentioned in chapter 2, I believe hog producers must avoid the temptation to become "drug and medication happy." Hogs and pork are considered in many corners to be overmedicated and the industry as a whole to be heavily drug dependent.

The truth is that there is at least partial truth to both charges, and this has hurt sales and consumption of pork and pork products.

In nature, when the population of a given species becomes too great for a certain locale, natural forces intercede to trim the population to more sustainable numbers. Generally these take the form of some sort of contagion or infectious agent. The same is true when domestic animals are packed too tightly into an artificial environment.

On farms with hundreds or thousands of head of hogs there is just no way to break disease and parasite life cycles. There are always little pigs, injured animals, stunted and stressed pigs, sows giving birth, and sick pens that never stand empty. These are never-ceasing reservoirs of sickness and infection.

I have an acquaintance who manages a 2,000-plus sow operation in Illinois in which many of his efforts and those of his 10-person staff go toward health care. Newborn pigs are actually given drug and vitamin infusions through stomach tubes. There are far too many swine farms now weighed down with such multiple-injection programs — not to increase performance, but to simply stay on line with at least some semblance of production.

Less May Be Better

I am neither an organic producer nor a health purist, but I do find that my workday and bottom line are both made better by breeding our hogs to be tougher and more durable rather than by buying animal drugs by the caselot. Producers with modest numbers can be truly discerning users of what is on the market. They don't have to jump on every new product down the pike just to stay ahead of the disease fires burning down in the hog lots.

The livestock industry appears to be on the verge of losing a number of current animal health products, has seen a number of others become available through prescription only, and may be shut out entirely from the use of any new-generation products that may also be usable by humans. This is because there are concerns about resistance developing to products if they are overused, and I think these concerns have some validity.

Due to the sheer numbers of animals — and health problems — on some farms, many producers now have to medicate constantly to keep those problems in check and reasonably manageable. Alas, their solution is being imposed upon a great many other producers whether they need it or not. How? Read on.

A lot of swine feeds are now laced with low-dosage antibiotics and other health products to aid animals in maintaining growth and performance in the presence of certain disease organisms. In fact, in many areas it is now all but impossible to buy complete pig starters and growers without low-dosage antibiotics. Such additions can and do increase the costs of feedstuffs.

Drugs and steroids have kept a lot of hogs in the gene pool producing, when they should have been washed out of the herd long before reaching the breeding pen. When these bad ones breed, so do their problems. Consumer confidence in pork will grow only as a commonsense approach to hog health care is applied all across the industry.

The counterpoint to subtherapeutic antibiotic use is management on a human scale to maximize the strengths and natural hardiness of the hog. In contrast to the large operator, small hog farmers have time to give to their animals, and that's what will ensure they are never driven out by the corporate operators. They can give attention to even the smallest details, can ever be about the business of fine-tuning the enterprise, and have a commitment to quality, harmony, and product wholesomeness that will always keep their production in demand.

Reading about many of the diseases reviewed later in this chapter will drill this point home. It will become apparent that many swine illnesses can be prevented through management — the type of management that the small producer can provide. Having said all this, I'll first share with you more specifics of my approach to good hog health care.

Five General Tips for Preventing Disease

Here are five ways that you can manage an operation for improved hog health:

1. Stagger farrowings. If you don't have too many little pigs around at the same time your workload will be reduced, enabling you to give attention to important animal health care details.
2. Practice lot and pasture rotation. This helps to control parasites and mud, which will keep the animals more comfortable.
3. Have empathy for hogs and provide extra time and care to those that require it. On most small farms every sow is known on sight, and her history comes quickly to mind.
4. Allow facilities to stand idle and open to the cleansing rays of the sun.
5. Employ selective breeding for vigorous and durable hogs. Select for deep, wide chests; large body cavities; high birth- and weaning-weights; large, loose frames; good mothering; well-formed genitalia; and obvious health and vigor.

Staying Current

To achieve good swine herd health, it is critical that you stay current with what is happening both locally and nationally. We have weathered many local disease outbreaks ranging from TGE (transmissible gastroenteritis) to pseudorabies in part by being plugged into the local scene. Through it you learn which local communities are experiencing disease outbreaks, which delivery trucks not to allow on the farm, which used-equipment sales not to attend, and sometimes when to just stay home, period.

Six different swine-oriented magazines arrive in our mailbox monthly, and we also receive a newsletter from our area's Extension Livestock Specialist.

Supply Kit

Every producer with hogs needs a medi-kit of supplies. I keep mine in a small plastic carrying tray grooved in the bottom to fit solidly on the top board of a gate or stall divider. This tray and its contents also fit neatly into a secondhand refrigerator we keep in an outbuilding to store animal health products at the correct temperatures.

The supplies my kit contains follow. You'll learn about the use of these products in the rest of this chapter:

❏ An assortment of disposable syringes in sizes from 3 to 10 cc's
❏ An assortment of needles in their sanitary plastic guards (used needles can be kept in a small screw-top jar filled with alcohol)
❏ A surgical blade handle and package of hooked and straight surgical blades (for tasks such as lancing abscesses)
❏ An aerosol wound protector or spray-type product (the latter does not have the hissing sound that many hogs find so upsetting)
❏ A marking crayon for identifying treated hogs (greens and yellows are the most visible on hogs)
❏ A rectal thermometer
❏ A broad-spectrum injectable antibiotic such as LA-200 (a form of Terramycin)
❏ An antibiotic product for use in the hogs' drinking water
❏ Epinephrine for shock
❏ Both a snare and a tongs for restraining hogs
❏ Prescription products as directed by the veterinarian
❏ A packet of a vitamin/electrolyte powder (which can be added to the drinking water to counter stress)
❏ An over-the-counter scours remedy
❏ A rehydration product
❏ A good standard animal health text such as *The Merck Veterinary Manual*

Veterinary Care

We have cultivated a relationship with our veterinarian that enables us to pick up the phone and get answers that still local rumors and give us first-hand information on the latest in animal health products.

To have such a relationship, you have to use your vet for more than midnight emergencies and salvaging animal wrecks. Our vet is on the farm several times each year for routine procedures such as blood testing, knows our herd and our management practices right down to the brands of feed we use, and knows me and my abilities. I also patronize him for a number of health supplies. For instance, he was the one who steered me to the speed syringe (which I'll discuss later).

Granted, his prices may be higher than those in some of the mail-order catalogs. Still, his products are always fresh and without close expiration dates, he can show me how to use them, they are available from him in small amounts, and they are just minutes away when needed.

Injections

We have raised literally hundreds of boars that received absolutely no other injections than their iron shot at birth. Still, injections are necessary to some extent, and they are one of those health care skills that even the smallest producer should acquire, to both trim operational costs and provide better care for his or her charges.

Besides the iron shot for newborns, you may have to provide antibiotics if your hogs experience a disease outbreak. Waterborne antibiotics are more appropriate for support therapy, but when you need a fast-acting antibiotic, you'll want to inject it.

The labels on the products you use for injection will advise you about the route of administration. Be sure to follow these instructions. These are three common injection methods for swine.

Subcutaneous, which means just beneath the animal's skin. Subcu shots usually are given in the shoulder/neck region.

Intramuscular, or into a muscle. This is by far the most common method of injection in hogs. With the other two methods — subcutaneous and intraperitoneal — absorption of the medicant is slower but more sustained. With intramuscular injection, there is rapid absorption, because muscles are well supplied with blood.

Intramuscular injections are given into large muscles that lie just beneath the skin. In times past, the prime target for IM shots in hogs was the ham muscle. However, there is an increased pain factor with injections in that region; some injections in the ham muscle have caused nerve damage and even lameness. The better target is in one of the long muscles on either side of the neck. These work for newborns right on up to the largest sows and boars. They're along the animal's topline, making them easily reachable; they're accessible even on animals held in a catchgate; and they're a good site choice for nearly all injectable health products.

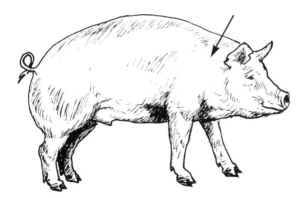

Common site for administering a subcutaneous or intramuscular injection.

Intraperitoneal, or within the peritoneal cavity, which is where loose connective tissue covers the surface of tendons and ligaments. Intraperitoneal injections go into hollows adjacent to body cavities. Perhaps most typical of this method is the injection of epinephrine to counter shock. On a shoat it can be given in one of the underleg areas that correspond to the human armpit or groin area.

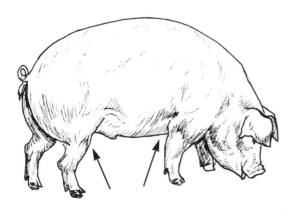

Sites where an intraperitoneal injection might be given.

How to Give an Injection

Giving an injection to an animal that's both larger than yourself and very agile can be a daunting task. One person who could not do it was my grandfather. He paid our local veterinarian for a field call to come to the farm to teach me how to give injections and do castration work. He also used to argue that the harder it was to inject a given hog, the less likely it was that that animal really needed an injection.

Our vet is able to get up to 20 cc's of an injectable product in a completely unrestrained sow with the following method:

◆ Use new needles to be assured that they are absolutely sharp and free of burrs. The few cents it costs to change needles is nearly always money well spent.

◆ Use one of the new disposable syringes or the speed models, made of nylon or plastic, that work with a squeeze of the hand. We have a 10 cc version of the latter that fits in the palm of the hand.

◆ Load the exact dosage needed, because the first injection is the only easy one.

◆ Use the shortest and heaviest-gauge needle you can — the thickest, in other words — or it may not penetrate the skin. To inject sows or boars I prefer a 1-inch, 14-gauge needle. For small pigs I prefer a ½-inch-long needle that is 18 gauge.

◆ Use a long-lasting product to reduce the number of injections needed for a full course of treatment. Some antibiotics now remain at strong levels in an animal's body for as long as 48 to 72 hours.

◆ Slowly and quietly approach the animal from the rear, holding the syringe parallel to the animal's topline.

◆ When the syringe is over the injection site, use a single motion to tilt the syringe up, drive the needle home, and depress the syringe.

◆ A pinch or hand slap on the site as the needle is withdrawn will prevent any serum leak from the puncture wound.

Personally, I have found that the two keys to making injections easier are the simple nylon syringes — which stroke smoothly and quickly — and clean, sharp needles of the shortest length and heaviest gauge possible.

As I have indicated, the giving of injections is a very useful skill, but if no amount of written instruction can give you the confidence you need to try it, enlist the help of your veterinarian. Chances are the vet will gladly give you a lesson, because if you can give injections in the future it will clear his or her calendar for more important diagnostic and health work.

Shots for Newborns

The following information appeared in chapter 8, but it's worth repeating if you're breeding hogs.

On their first day of life I give newborn pigs a ½ cc injection of a long-lasting antibiotic, to counter any potential problems with navel ill or other early infections. The injection site is one of the long muscles in the side of the neck.

In the other side of the neck I give a 1 to 2 cc supplemental iron injection, to prevent iron deficiency anemia. (See "Iron Deficiency" on page 216.) I follow this with another 1 cc injection of iron when the pigs are 10 to 14 days of age. A third iron injection of 1 cc may be in order at weaning or shortly before.

There are alternatives to injectable iron that can be given orally or in a group of products commonly called "rootin' iron." The latter come in a peat moss mixture that is spread around so the pigs actually root in it. I prefer the injectables, followed by the oral products; these ensure that each pig gets an exact amount of the needed iron. For the "needle shy" the oral pump bottle may be appropriate.

Parasite Control

Hogs contract a variety of parasites, which can adversely affect weight gain and, in some cases, make pigs sick. Some parasites, such as roundworms, are internal; others are external parasites, such as mites, which affect the skin. For more information on specific types of parasites, see "Swine Ailments and Illnesses," beginning on page 209.

Early methods of countering internal parasite buildup, or worms, depended almost totally on regular rotation of various swine lots and pastures to break up worm life cycles; each lot was used just once a year. Some rather noxious and dubious concoctions were also fed or given in the drinking water to try to knock out worms. Today, we have more effective options for worm control.

Still, no one product seems up to the task of removing all of the parasites that can affect hogs. That's why it's important to rotate the parasite control products you use. Further, there is some strong evidence that repeated use of the same worming product may cause some parasite species to develop a resistance to it.

Types of Wormers

In no way do I wish to appear to favor one product over another here. Many worming products are useful, even some of the older ones such as Piperazine. Still, at Willow Valley we rotate the products Ivomec, Atgard, and Tramisol.

With Ivomec, which contains the anthelmintic ivermectin, we control not only a wide variety of internal parasites but external parasites such as mange mites and lice as well. Ivomec comes in an injectable form as well as a pour-on formulation.

Atgard kills many internal parasites at all stages of development. It is a granulated product that must be mixed with mealtime rations, and should not be mixed with pellets. You can buy it premixed.

Tramisol is a powder that is mixed with water, then added to drinking water. It kills lungworms, which rely on red earthworms as an intermediate host and can be a special problem in hogs living outside or in wooded lots.

Administering Parasite-Control Products

As noted, some worming products can be given by injection, others can be mixed into feed (treated feed is the only ration offered to the hogs over a period of from one to three days), and still others are mixed into drinking water. Products for external parasites, such as Ivomec, may be injected, but they also come in other forms, such as pour-ons. The surest way to worm is by injection, because you are ensured that each animal gets the exact dosage for its weight and age.

It is more difficult to administer wormers in water, especially during cold weather, when water usage is reduced and there is the risk of the water freezing before all of the wormer is consumed. And if you administer wormers in the feed, there is always the concern that the smaller pigs in the group or pen won't get enough feed and, thus, enough wormer.

For control of the more common external parasites, lice and mange mites, you can choose the injectable Ivomec or any of a number of spray, pour-on, and dust-type products. Your selection will be dictated by both the environment and the season of the year.

The sprays and pour-ons are most effective for external parasites, but they have two drawbacks: Some cannot be used too close to farrowing, lest traces of the product be ingested by the nursing pigs; also, in weather below 40°F problems with chilling and stress may arise from the wetting of the hogs.

Lice problems increase in cold weather, when the hogs lie up more in the bedding and sleep closer together. A good pour-on can be applied along each hog's topline on a pretty fall day, and dust-type products can be used on the hogs and bedding in cold weather. The dust-type products may be among the least effective, however, due to reduced contact times; still, they may be help-ful in the coldest of weather.

Be aware that there is a withdrawal time for some parasite-control prod-ucts. You cannot use these products for a certain number of days before the hogs go to market; for some products, it's as long as 75 days. Be sure to read the label on each product to determine its withdrawal time.

Parasite-Control Schedule

It's a good idea to check with knowledgeable hog producers in your area or your local swine veterinarian before selecting parasite-control prod-ucts. These people can help you determine which products will be most effective for the parasites you need to control. However, you do want to rotate worming products to help ensure that parasites don't become resistant to one specific kind. Try not to use the same worming product more than twice in a row.

Here is an example of a parasite-control plan based on the way we do things at Willow Valley.

New pigs brought onto the farm. Seven to 10 days after arrival, treat with injectable Ivomec or Tramisol. Then administer a second treat-ment of Ivomec or Atgard when the animals reach 150 to 185 pounds.

In the breeding herd. Treat all breeding animals for external para-sites. If it's warm weather, a liquid pour-on may be best; if it's cold, a powder. Injectable ivermectin is a good option but may be costly.

Sows. Two to four weeks before farrowing treat sows for internals and externals. If you did not use injectable Ivomec last time, it would be a good option now. Otherwise, you might want to treat this time with Tramisol and use something else for external parasites, such as one of the pour-on or powder products.

Piglets. We worm at six to eight weeks, about one week after wean-ing, with Ivomec injectable.

Alternative Parasite-Control Products

An organic parasite-control method is the feeding of and dusting with diatomaceous earth. This is a component of a product that we administer to our hogs called Immuno-Boost. When we offer this product in the drinking water along with another worming measure, it does seem to increase the hogs' response to the wormer.

I have never used exclusively diatomaceous earth for worming and have doubts about just how effective it can be, but I do think it gives some results. Anyone into organic hog farming should also be diligent about rotating pastures to help control parasites.

Swine Ailments and Illnesses

One chapter in a book cannot begin to cover all of the illnesses and ailments that affect swine, but here you will get some of the basics.

There are a number of problems in swine that you'll be able to manage yourself; for others, you'll need to enlist the help of your veterinarian. Most important, however, is that you familiarize yourself with the signs of various important illnesses in your hogs. If you have any doubt about what's wrong, or that you can handle the problem yourself, consult your veterinarian. In hogs, prompt treatment of diseases can be life saving. It can also prevent reductions in weight gain.

Abscesses

These are puss-filled swellings caused by bacterial infections. They are most noticeable when right under the skin and can range from thumb to fist size. Some hog producers lance the abscess with a clean blade and apply hand pressure to facilitate drainage, but this procedure can introduce the abscess-causing organism to the rest of the animals in a pen. If you lance, do it away from other animals. Afterward, apply a good common drying astringent, such as Blue Lotion or an iodine product.

Treatment from "within" is the more common approach. Abscesses normally respond to the more commonly used injectable antibiotics.

Atrophic Rhinitis

This is a mouthful of a disease generally just called rhinitis. It is a respiratory disease that is thought to be due to infection with more than one "bug." It is transmitted to young pigs probably via nose-to-nose contact with their dam.

Atrophic rhinitis results in inflammation in the mucosal lining of the nose. Early symptoms include sneezing, sniffling, possibly a bloody discharge from the nose, and a watery discharge from the eyes. In more advanced stages the bones in the snout (turbinates) may atrophy, and the snout may even swell and twist. The disease can dramatically slow growth, and as turbinate damage progresses the hog also becomes far more vulnerable to pneumonia.

In the advanced stages of atrophic rhinitis, the bones in the snout (turbinates) may atrophy, and the snout may even swell and twist.

In the very early stages of disease, rhinitis may go undetected in white breeds; the eye and nose discharge may not be as noticeable on white faces as on dark.

There is a preventive injection — a vaccine — that can be given to sows before farrowing, and to young pigs in herds with this disease. It is usually combined with antibiotic treatment and changes in management procedures. However, once the vaccination program for this disease is initiated on a farm, it must be continued.

At one time it was thought that as many as 85 percent of the hogs in the United States had at least a touch of atrophic rhinitis. Animals sure to be free of rhinitis and some other diseases are those taken from their dams by cesarean section and raised in isolation; they are called SPF ("specific pathogen free") pigs.

Although some of the symptoms are similar, this is not the same disease as necrotic rhinitis or bull nose in young hogs.

Vaccinations

With good selection, care, and feeding, I have found that vaccinations against most respiratory diseases are seldom necessary. The problems we've had were the kind that, generally, we could treat as needed. However, some herds develop persistent diseases for which vaccination may be beneficial. If you have a herd that develops an infectious disease, discuss a vaccination program with your veterinarian.

Brucellosis

This is a bacterial disease that is declining in significance in the United States following many years of blood testing breeding herds and certifying them free of the disease before breeding stock from them is allowed to be sold.

Its most common symptoms — which can cause great economic loss — are abortions, stillbirths, and the birth of very weak pigs that fail to thrive. Brucellosis also causes an infection of the testicles in males and can lead to infertility. When carriers of this disease are found by testing they are ordered destroyed. Undulant fever is the human form of brucellosis, and it once took its toll on farmers and animal health workers.

Clostridial Enteritis

This is a bacterial gastrointestinal disease usually caused by an organism called *Clostridium perfringens* Type C. Veterinarians, especially in the Midwest, have reported an increase in the incidence of this disease in recent years.

Clostridial enteritis is usually seen in pigs' first few days of life, but it can affect weaned pigs, too. It can cause diarrhea that may or may not be bloody. Sometimes the disease is so virulent it causes sudden death in piglets before you even have a chance to realize anything is wrong. It is thought that the pigs contract the disease from the sow and from sow feces; the organism is also found in soil.

Clostridial enteritis can be very difficult to distinguish from other gastrointestinal diseases, especially those caused by another organism known as *E. coli*.

The treatment for clostridial enteritis is generally antibiotic treatment; injectable and feed-grade antibiotics also are often given, and veterinarians often advise management procedures, especially beefed-up sanitation efforts aimed at reducing the spread of disease to piglets. This would include washing the sow before farrowing. Antitoxins and vaccines also have been used. It can be difficult to get clostridial enteritis under control once a herd is infected.

A less severe, chronic, "Type A" form of this disease has also been identified. Infested pigs might develop mild diarrhea, but not always, and the Type A disease can cause reductions in weight gain and feed efficiency. At least one veterinarian has reported that the Type A form has, in some herds, increased preweaning deaths among pigs that develop diarrhea; in pigs that survive, it can reduce weaning weights. Treatment may be necessary if the disease is causing problems, but in some herds this milder form of clostridial enteritis clears up on its own.

Dehydration

Dehydration is usually caused by illnesses that result in scouring, or diarrhea. Oral glucose and electrolytes are administered as antidotes for dehydration, and there are a number of good ones such as Re-Sorb, available just for pigs. In a pinch you can use Pedialyte, a product for children available at your local drugstore, or even Gatorade, which can be administered orally with a syringe (without the needle attached). Keeping dehydrated animals warm and comfortable is also a good idea. The real trick is probably to keep health problems away from the farrowing units and the breeding herd.

Signs of Illness

Throughout this book you have learned about the signs of illness in swine. Here is a roundup of 10 of them; all indicate you could have an infectious disease problem. Suspect trouble if animals are:

1. Slow getting off their beds
2. Eating less or eating not at all
3. Looking drawn
4. Developing diarrhea
5. Acting depressed
6. Vomiting
7. Experiencing abortions or stillbirths
8. Lame or walking stiffly
9. Acting uncoordinated
10. Among nursing pigs, seeming to be "poor doers"

Erysipelas

This is an old-line disease that still pops up from time to time to take a toll on hogs. It is a bacterial disease with symptoms ranging from loss of appetite and fever, to a stiff gait and arched back, to the classic red, diamond-shaped patterns appearing on the skin, to sudden death with no apparent symptoms.

Both injectables and a waterborne product are used to treat the disease.

E. coli

This is a bacterium that can cause several illnesses. Diarrhea after weaning, for instance, known as postweaning scours, has been associated with *E. coli* (although this problem has also been linked to early weaning). One of the more serious conditions linked to *E. coli* is hemorrhagic bowel syndrome,

Hog Cholera Eradicated

Hog cholera used to be dreaded in swine. It was a highly infectious viral disease that came on suddenly and had a high death rate. But thanks to an eradication program initiated in this country in 1962, hog cholera has been wiped out. It is still a problem in several other countries.

which is a highly fatal, massive hemorrhage into the intestines. If your pigs break out with any serious infectious disease, consult your veterinarian immediately, to determine the cause and find the right treatment.

Gastric Ulcers

One cause of this problem is thought to be giving hogs feed that is too finely ground, although factors such as stress might also play a role. Hogs with gastric ulcers may act dull and go off their feed. If their ulcers bleed, they may become anemic.

To prevent ulcers, be sure to give your hogs feed that is more than 3.5 mm screen size, and keep them comfortable to reduce stress.

Glasser's Disease

The clinical name for Glasser's disease is Haemophilus polyserositis. That's because it's caused by the organism *Haemophilus parasuis* (*H. suis* for short), and it involves the joints as well as a generalized illness. The *H. suis* organism is one that hogs can harbor in their respiratory tracts without experiencing any problems; however, during times of stress, such as weaning or temperature changes, it can make pigs sick.

Glasser's disease develops suddenly and kills pigs quickly. It tends to strike pigs under 10 weeks of age. There is also a chronic form of the disease that causes all kinds of symptoms, including a fever, trouble breathing, swelling in the joints, and lameness.

This is a disease that tends to be more severe in hogs that are specific pathogen free (SPF); those are pigs delivered by cesarean section and free of many diseases, such as mange. However, when SPF pigs are introduced into conventional herds and have no immunity to *H. suis*, they can become very ill with Glasser's disease at any age.

Treatment needs to be administered as quickly as possible and consists primarily of antibiotics. There is also a vaccine which a veterinarian can recommend if *H. suis* becomes a problem in your herd.

Greasy Pig Disease

This is nothing more than a staph, or *Staphylococcus*, infection. If pigs have external parasites, such as mange, or traumatic injuries to the skin that are not treated, the infection can set in. Improper nutrition is also thought to be linked with the disease.

Greasy pig disease tends to affect pigs either just before or right after weaning. The skin flakes off, becomes crusty, and may crack. There may be a stinky dark brown secretion. Pigs may act dull and go off their feed. The disease can kill pigs if not treated.

Hyperthermia and Hypothermia

Hyperthermia means the body is overheated. I once lost a good gilt to hyperthermia in the middle of a snowstorm. A border collie puppy chased her around until she got overheated.

A hog that gets hyperthermia goes into shock. Do whatever you can to calm it. Get it away from anything that might upset it. *Do not hose the hog down,* however. If you cool down a hot hog, you'll have a dead hog on your hands. At most, brush a bit of cool water across the bridge of the nose. Administer an intraperitoneal injection of epinephrine, and keep your fingers crossed.

Hypothermia, or low body temperature, means the animal is too cold. It isn't a common problem in hogs. Little pigs, however, are susceptible to chilling, and keeping them warm is important. Be sure to read up on ways to do this in chapter 8.

The remedies are antibiotics, washing the pigs with medicated shampoos, and, sometimes, dipping them in solutions containing products such as Chlorox. To hasten healing, pigs should be treated for external parasites if they have them, and measures taken to see that their environment is comfortable and their feed adequate in nutrients.

Taking the Temperature

The chart below, adapted from the text *Diseases of Swine,* 7th Edition (Iowa State University Press, 1992), shows the normal temperature for pigs of different ages.

In many other types of animals, it's common to check the heart and respiratory rate as well as the temperature. But in pigs and hogs, I've found that the temperature, coupled with the way the animals act, is a pretty good indicator of illness.

What does the temperature mean? As you'll see below, the hog's normal body temperature is about 101 to 102°F. When a fever develops, it's because the body's systems are rising to its own defense to fight off an infection — which it may or may not be able to do. But a below-normal temperature can indicate trouble, too. It can mean the body systems are shutting down, which occurs with kidney failure in swine.

When I take a hog's temperature, I use a standard rectal thermometer with a ring in the end. With a short piece of stainless wire I attach an alligator clip, which I then attach to the animal's haircoat. (This leaves my hands free for other chores while I wait.) I insert the thermometer about 3 inches into the rectum and leave it in for at least two minutes.

Pig Age	Rectal Temp.
Newborn	10.2° F
1 hr. old	98.3° F
24 hrs. old	101.5° F
Unweaned pig	102.6° F
Growing pig (60–100 lbs)	102.3° F
Finishing pig (100–200 lbs)	101.8° F
Gestating sow	101.7° F
Boar	101.1° F

Infectious Arthritis

This is an infection that strikes the joints. Lameness is the most obvious sign of infectious arthritis in the pig; it may affect only one limb, but it can also affect more. In newborn pigs, it may be due to exposure to pathogens in the uterus; after birth, it can result from navel ill, or pigs might contract infections during tail docking, castration, and other management procedures. A number of organisms may be the culprit, including *E. coli*, *Streptococcus*, and *Staphylococcus*.

In slightly older pigs, or pigs over about four weeks of age, infectious arthritis may be due to an organism called *Mycoplasma hyosynoviae*. This is an organism found in the nose, which then moves to the joints.

Some forms of infectious arthritis respond better to antibiotics than others, and some medications only work if given early in the course of disease, so you'll need a veterinarian's help to determine the exact cause of and best treatment for the problem. Avoiding the conditions that result in infectious arthritis is a better approach. This could be concrete flooring, kneeling at feeding equipment, or any other situation that causes damage to knees and hocks.

Iron Deficiency

I've advised more than once in this book that supplemental iron be given to baby pigs. Without it they are at risk for developing baby piglet anemia, or iron deficiency, because baby pigs require more iron than they receive from their sows. Baby pigs can become very anemic within a few days of birth.

Pigs with anemia won't grow normally, and they act dull. Their skin is pale and they may scour or even have trouble breathing. They may also have a longer-than-normal haircoat.

If the initial dose of supplemental iron is injected, rather than orally administered, you can be sure the pigs receive what they need.

Leptospirosis

Commonly known as "lepto," this disease is most often caused in swine by the infectious organism *Leptospirosis pomona*. The organism can be transmitted through the urine of infected animals, which may be dogs, cows, or rats as well as swine. It enters the body through the mucous membranes or breaks in the skin. Lepto can cause a fever, loss of appetite, and diarrhea. But its most common and costly effects in swine are abortion, stillbirths, and the birth of weak pigs.

Antibiotic therapy under the direction of a veterinarian is the recommended course of action after the cause of any abortions has been determined. There's also a vaccination for lepto, but the immunity it provides may be short lived.

Lice and Mange

These are common external parasites of swine.

The pig louse *(Haematopinus suis)* is more likely to be seen in herds that are not well managed. It is transmitted via pig-to-pig contact. The lice actually suck blood from the pig's skin, which can result in anemia. The primary problem lice cause, however, is skin irritation and itching. Infested pigs may grow slowly and don't efficiently utilize their feed.

Suspect lice if your pigs rub themselves to relieve itching, and especially if you see small white eggs on their hairs — which are lice nits. You can see lice on a pig, too, around its head, ears, and flank area. The remedy is use of a product that kills lice.

Like lice, mange is not uncommon among swine. It is an infection with a mite called *Sarcoptes scabiei*. This external parasite spreads via pig-to-pig contact, and it can be spread from an infected dam to her litter.

Mites favor the head and ear area. Hogs with mites will rub to relieve the itching, and may develop scabs. Some young pigs are allergic to mites, so their response to this infection will be more severe. Their itching will be worse, and skin irritation can involve their hind ends and abdominal areas. Young pigs with mite infestations grow more slowly.

The remedy for mange mites, as for lice, is treatment with a product that kills the parasites. Be sure to read "Parasite Control," beginning on page 206.

louse

mange mite

Lungworm

Hogs contract this parasite by ingesting infected earthworms. Larvae migrate to the heart as well as the lungs.

The result of lungworm infestation is similar to that of some stomach worms: Hogs may have trouble breathing and an asthmatic cough characterized by a "thumping" noise. They may be "poor doers," animals that are growing slowly.

Good sanitation and an effective parasite-control program are the ways to control lungworm.

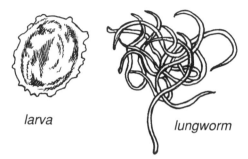

larva

lungworm

MMA (Mastitis, Metritis, and Agalactia)

This is a syndrome characterized by mastitis (udder infection), metritis (uterine infection), and agalactia (loss of milk complex). It can strike sows shortly before or after farrowing. Sows may lose appetite, have a discharge from the vulva, run a fever, appear depressed, have a hardening of the udder, or even dry up completely. In short, a sow isn't able to provide enough milk for her litter. She may also become constipated; and her mammary glands may be enlarged and feel warm.

This syndrome can be caused by several problems, which include infections, stress, and dietary management problems. It is more common in sows that have had more litters.

Sows that develop MMA may require antibiotics and oxytocin to stimulate milk flow. Something to encourage appetite might also be in order: You might try pig starter or gelatin powder to stimulate their curiosity. There is also a product called Spur intended for this. Vitamin B injections may do the trick, and we've had good results by administering LA-200, an injectable antibiotic.

I've heard about giving sows everything from dog food to beer to stimulate appetite, but my concern there would be that such a change in diet might cause more problems than it resolved.

The best approach to MMA is prevention. Add bulk to the sow's diet, keep her comfortable and clean, and provide her with opportunities to exercise.

Tracking Treatments

When you are administering treatments, such as antibiotic injections, to several animals it can be difficult to keep track of which ones you've treated and which you haven't. To resolve this problem, I mark each animal as I administer the treatment. If some animals require more than one dose, I use two different marker colors: for instance, a red marker as I administer the first treatment, a green marker for the second, and so on.

For little pigs, I've found that marking on the top of the head ensures that I see my color coding.

Mycoplasmal Pneumonia

This is another mouthful of a disease, and it's caused by an organism called *Mycoplasma hyopneumoniae*. It is thought by some to be the most common cause of chronic pneumonia in swine. It can be spread in the air and via direct contact.

This form of pneumonia is more likely to be seen in pigs at least three months of age. They develop a chronic, dry cough, which might not be evident until they exercise. They may or may not have a fever or lose their appetite, depending on the seriousness of the infection. This disease, like many others that affect swine, can stunt growth.

Mycoplasmal pneumonia may clear up on its own if pigs haven't been stricken too severely, but the problem is that when pigs have this disease, it often makes them susceptible to other infections, such as *Streptococcus suis*. In such cases antibiotic treatment is necessary; there is also a vaccine available. This disease, too, is associated with overcrowding and poor sanitation, so avoid such stressful conditions.

Navel Ill

This is a general term used to describe an infection that starts in the umbilical area and spreads. The navel is one of the primary routes of infection into the very young pig. Infections that originate at the navel can be prevented by applying an astringent, such as an iodine solution, wound spray, or Blue Lotion from 12 to 24 hours after birth.

In pigs that develop navel ill, antibiotic treatment is probably necessary.

Necrotic Rhinitis (Bullnose)

This is an uncommon disease, and I mention it here only to make it clear that necrotic rhinitis is different from atrophic rhinitis. Necrotic rhinitis is a bacterial infection that generally occurs on farms with poor management and sanitation. It might develop if management tasks such as nose ringing or wolf teeth cutting are performed under less-than-hygienic conditions, or after young pigs cut their wolf teeth. It's obvious where the common name of the disease — bullnose — came from: Infected pigs develop serious swelling and abscesses of the snout. Improved sanitation practices and antibiotic treatment are used to resolve the condition.

Porcine Reproductive and Respiratory Syndrome (PRRS)

This is a viral disease that results in several problems affecting — as the name implies — the reproductive and respiratory systems. It is thought to be spread via pig-to-pig contact, but airborne transmission is also possible. PRRS is considered one of the newer diseases, first reported in the late 1980s. Initial outbreaks occurred in Indiana, Iowa, and Minnesota.

Swine affected by PRRS may develop a fever, lose their appetite, and appear dull. The respiratory part of the syndrome causes the classic "thumping" type of breathing, and sets the stage for bacterial infections to take hold. Sometimes PRRS results in a lack of oxygen that turns the ears blue, a condition called blue ear disease. Abortions are another significant problem linked with the disease. PRRS tends to clear up after about two months, but can sometimes be fatal.

The government has been working with hog producers whose herds are affected by PRRS to study the disease by taking blood and tissue samples. Currently there is no specific treatment for PRRS, and the best way to avoid it is to prevent your hogs from exposure to herds that have been sick with the syndrome. PRRS may also be more likely to occur in herds that have been exposed to fumonisin mycotoxin, which is a toxin formed from a fungus that can occur in feed.

Pseudorabies (Mad Itch)

This infection is caused by a herpes virus, and it is the current hot disease in swine circles. Most states now mandate a testing program for producers who sell breeding stock, to help hold it in check. It can affect hogs of all ages, but quite often first appears as abortions and weak pigs at birth. Weaned and growing pigs with the disease develop pneumonia, a fever, poor coordination, and convulsions. Pseudorabies is usually fatal in pigs that aren't yet weaned; it's somewhat less deadly in weaned and growing pigs. Adult hogs with pseudorabies may have pneumonia or fever — or show no signs of infection at all.

Other farm animals as well as dogs and cats can contract the disease. All tend to experience intense itching, hence the name mad itch. Pseudorabies is virtually always fatal in these other species.

There is no treatment for this disease, although antibiotics are sometimes used to treat the resulting pneumonia.

Extensive testing and destruction of infected animals is under way to control this costly disease. There is an injectable preventive, but although it may reduce clinical signs of the disease, it may not prevent the spread of infection in the herd. There's another problem with this product, too: Once it is used, your animals will test falsely positive for pseudorabies, and that can be murder if you are in the breeding business. I thus recommend not using the preventive unless the problem with pseudorabies is really serious in the herd.

Rotaviral Enteritis

This is a virus that is thought to affect many herds. It may result in diarrhea among newly weaned pigs, and it's one of those diseases to suspect when nursing pigs are "poor doers." Rotavirus is often found along with other swine infections, such as E. coli.

There are vaccines available, but their results have been mixed, because they may not tackle all of the strains of rotavirus that are affecting a herd. This disease requires good management of weaned pigs by providing them with a stress-free, comfortable environment and good diet. Treatment of bacterial infections that might occur concurrently with rotavirus may be needed, as well as nursing care, such as rehydration of pigs with rotavirus that have diarrhea.

Salmonellosis

The disease salmonellosis results from infection with the bacterium *Salmonella*, which can be spread a variety of ways. Pigs can contract the infection from other infected pigs, or by coming into contact with infected feces. They can also contract *Salmonella* if their feed is contaminated. A USDA/APHIS study of feed samples from 300 finisher operations showed that the incidence of *Salmonella* was less than 6 percent. Of the operations in which the primary grain was corn, 5 percent were positive for *Salmonella*, and of those that fed some other grain besides corn, 20 percent contained this pathogen.

There are different types of *Salmonella* "bugs," and they cause different diseases. Thought to be the most common in swine is the one known as *Salmonella choleraesuis,* which can lead to pneumonia, gut problems, and septicemia, which means the infection spreads throughout the body.

If the respiratory system is affected, swine may cough and have trouble breathing. If their gut is affected, they develop diarrhea, which can come and go and, of course, lead to dehydration. When salmonellosis becomes septic, other internal organs besides the lungs are affected, such as the liver and spleen. Other signs of the disease include loss of appetite, fever, and depression.

Swine with this disease need treatment with an antibiotic, and you'll need the help of your veterinarian to determine exactly which one will work. Salmonellosis has been linked with poor sanitation, overcrowding, and bringing in feeder pigs from multiple sources, so heed advice about keeping pigs clean and comfortable and buying stock from one source whenever possible.

Salt Poisoning

Pigs can develop salt poisoning by consuming excess salt in their diet, but more commonly the problem begins when they are deprived of water, then drink plentifully. Salts accumulate in the body with water deprivation, and when the animal drinks, water is drawn into the central nervous system. Swelling in the brain results.

The signs of salt poisoning include restlessness, constipation, thirst, and depression. Salt poisoning can lead to blindness, deafness, convulsions, and even death.

Hogs can be inadvertently deprived of water if their source freezes, if there is not enough space around the waterers for the number of animals you have, or if your automatic waterer stops working. If you have to give pigs medication in water, they may not like the taste and stop drinking. Ensuring that your pigs have access to water is very important. If the problem is medicated water, try adding flavored gelatin to mask the unpleasant taste.

Watch the Salt

Generally, the salt content in feed should be less than 1 percent.

Scouring

Scouring, or diarrhea, can be caused by a variety of factors, ranging from dietary problems to infectious disease. It can be lethal in a very short time, and unless you are very sure of the cause, consult a veterinarian to determine that cause as well as a course of action. Prompt treatment will not only increase pig survival but also help maintain reasonable growth in the face of a health problem. Pigs with scours usually become dehydrated and need rehydration. See "Dehydration" on page 212.

Shock

Shock occurs when there is inadequate blood flow to the body tissues. There are numerous causes, including severe stress, trauma, serious infections, and dehydration. I've found that trauma is the most common cause.

Swine in shock may be prostrate and have a rapid pulse, rapid breathing, and low body temperature. Hogs in shock need to be kept warm and dry, and should receive treatment for dehydration. I also administer an intraperitoneal injection of epinephrine *immediately* if I suspect a hog is in shock. It usually works fast. The cause of shock also needs to be addressed — wounds must be treated, for example.

Splay Leg

This is a condition of piglets in which their legs are splayed out from the body. The defect will be readily apparent soon after birth. Hogs can have a genetic tendency toward this condition, which underscores the importance of selecting hogs carefully for breeding.

Another cause of splay leg can be slippery floors, so provide good footing for your animals.

Splayed legs are readily apparent.

Stomach Worms

There are several types of internal parasites that occur in the stomachs of swine. Among them are the red stomach worm and the large roundworm.

The red stomach worm is likely to be seen in hogs raised on pasture. The only sign of this parasitic infection may be less-than-optimum weight gain, but this parasitic infection can also result in stomach irritation and ulcers.

The large roundworm is more common. Indeed, the females are prolific — they can lay over a million eggs per day. The eggs can be transported on your boots and are pretty resistant to disinfectants, although heat or sun will render them unviable.

Growing pigs are more often affected by the large roundworm than are adults. Also, as with the red stomach worm, this parasitic infection is more common in swine raised on pasture. Although considered a stomach worm, the large roundworm migrates to the lungs, and that's where it can cause problems. Infested pigs may have trouble breathing and an asthmatic cough characterized by a "thumping."

An effective worming program is the treatment for stomach worms, as well as improved sanitation.

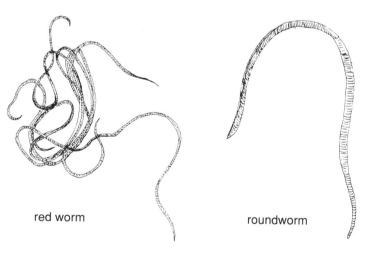

red worm   roundworm

Needle Disposal

Safe disposal of needles is a concern; in some areas there are regulations governing it. Generally, it is advised that the needle be broken off and inserted into the syringe before disposal, but it can be difficult to break off some of the heavier needles used for swine. Check with your veterinarian for further advice about how to best dispose of used needles.

Swine Dysentery

This is an infectious disease that is usually seen in growing and finishing hogs. It is spread by infested feces not just from pigs but also from birds, rats, and mice. Although most hogs recover, the death rate from dysentery can be as high as 30 percent.

When dysentery starts you may see yellow- or gray-colored feces, then mucus and blood in the feces. Hogs with dysentery may or may not develop a fever or go off their feed.

Antibiotics administered in food and water generally are used to destroy the organism that causes swine dysentery. There is a vaccine available, but its results have been mixed. Good sanitation, such as providing hogs with clean housing and the elimination of any rodents, will help control the disease.

It can be hard to tell dysentery from other diseases that affect swine, so you'll need your veterinarian's help in diagnosing and treating the disease.

Swine Influenza

The hog version of the flu, this is usually caused by the Type A influenza virus. It is a disease that can be transmitted both among pigs and between pigs and humans. It can also be spread to pigs via lungworm eggs, which are passed in the feces of infected pigs then eaten by earthworms, which are in turn eaten by the pigs.

The signs of this disease in pigs are similar to those you would have if you came down with the flu: It comes on suddenly and causes a deep, dry cough, fever, and loss of appetite. The animals may have a discharge from the eyes and nose. Hogs usually recover from the flu, but what often complicates the picture is a secondary bacterial infection. In addition, sows that are infected during pregnancy can have small, sickly litters.

There is no specific treatment other than good nursing care. Pigs should be kept comfortable and under minimal stress. They might need antibiotics in their drinking water if they also develop a secondary bacterial infection.

Swine flu is often brought onto a farm when new animals — feeder pigs or breeding stock — are introduced. It tends to occur in the fall and winter months, and after there have been wide fluctuations in temperatures, which stress pigs. Some pigs carry the disease even though they have no signs of illness. That's why it's so important to shop carefully for new pigs, make sure they're disease free, and do everything else possible to prevent exposure of your hogs to anyone or anything that might introduce disease.

Disease of the Month?

So often I hear hog producers in a panic over the "disease of the month." More often than not, these tend to be diseases that are affecting commercial hog producers with large numbers of swine in confinement operations. This or that disease gets a lot of press, and pretty soon everyone is alarmed. These large commercial operations have a continuous flow of pigs in and out, which is more conducive to the spread of certain diseases than is the environment on small farms.

Of course, anything is possible, but if you select your animals for hardiness and manage your farm well, it's unlikely you'll have too much trouble. You can further protect your animals by keeping them away from perimeter fences, and by prohibiting feed trucks anywhere near your hog lot. The undercarriage of a vehicle can carry disease organisms. If you know there's an outbreak of something like TGE in your area, stay home to have your cup of coffee instead of having it in the town diner.

Transmissible Gastroenteritis (TGE)

This is a serious, highly contagious disease caused by the coronavirus, and it's one of the more important diseases around today. It can affect pigs of all ages, but young pigs are the ones that become most ill.

TGE tends to occur in winter months. Pigs get it when they ingest the virus, which they can get from many sources: infected pigs, fecal contamination of your boots as you walk into the hog pens, and even dogs and birds.

Little pigs with TGE experience vomiting and diarrhea, become dehydrated, and often die, especially if they are only a couple of weeks old or younger. Older pigs also have vomiting and diarrhea, and they may go off their feed; sows might develop a fever. However, older pigs often recover from the disease.

A milder, chronic form of this disease also occurs that is thought to be due to more than one "bug."

There's no specific, effective treatment for the virus that causes acute TGE. Good nursing care is crucial for pigs with acute TGE; they should be treated for dehydration and kept in a warm, dry place. In the chronic form of the disease, other "bugs" may be involved, so antibiotic treatment may be indicated.

To avoid TGE in your herd, be careful not to buy or bring in infected animals. Don't go to other places where there are hogs, or hog producers, in the same clothing you wear around your hogs. Keep birds and dogs away from your herd.

Trichinellosis

This isn't nearly the problem people believe it to be — the fear really is needless. Trichinellosis in swine is a parasitic infection. It can be transmitted via undercooked meat to humans, in whom the disease is called trichinosis.

Improved swine management and increased public awareness of the importance of thoroughly cooking meat have effectively eliminated trichinosis. In fact, the last I heard about the disease in people, it was associated with eating bear meat, not pork.

Whipworms

If you have pigs from about two to six months of age that develop bloody diarrhea with mucus in it, suspect whipworms. Whipworms are spread by whipworm eggs in feces, which are then ingested by pigs.

An infection with whipworms, also known as *Trichuris suis*, leads to dehydration and wasting. Pigs with a whipworm infection may also go off their feed and become anemic. They can die.

As with most types of parasitic infections that affect swine, good sanitation and an effective parasite control will control this problem.

Treating Wounds

In hogs, the most common wounds are scrapes, cuts, and punctures. They tend to be shallow and will often heal with little or even no treatment. But an aerosol wound spray, such as Blue Lotion, or tamed iodine can be applied quickly, disperses over the wound site, dries quickly, and can be used without restraining the hog.

Some hogs become skittish when they hear the hissing sound of an aerosol. Fortunately, many wound sprays are now available in pump and squeeze bottles. Our veterinarian even packages a treatment he prepares for wounds and to treat navels into squirt bottles, which are available in several sizes.

With very deep puncture wounds, however, some products may not penetrate fully. You run the risk that the wound will heal over at the surface but leave an infection underneath. For such wounds keep on hand a prescription product made for this purpose that you can obtain from your veterinarian; it's more potent than what you'll buy over the counter.

Just be careful when using any spray around a hog's head not to get the product into the eyes or ear canals. Watch out for your own comfort, too. Tamed iodine is supposed to be milder than full-strength iodine, but it still always seems to find its way into that one little nick you have on your hand; you'll find that the word *tamed* is a subjective one!

TEN

MANAGING YOUR HOG OPERATION

WHETHER YOUR SWINE venture numbers 2 head, 200 head, or even 2,000 head, it is very much a business and should be operated as such.

To be successful with hogs you have to do two things: get muddy and get serious about making a profit. And to achieve the latter you have to know where every hog in the herd is, how it is performing, and how much it is costing you.

Record Keeping

Experience teaches that the best-looking and best-doing animals in the herd are seldom — if ever — the same ones. The only way to find those best-performing individuals lies in maintaining detailed records on herd performance.

We once owned a Hampshire sow that we used to produce F_1 Duroc x Hampshire boars and gilts. Her first litter was born early in a hot August and, with her pigs at a bit over three weeks in age, she went off her feed and began to suffer heat stress. For nearly 48 hours she didn't eat — but she did continue to nurse her pigs. The toll it took upon her was visible all of her life.

At a time when she needed food to make needed growth as a first-litter gilt and nurse her pigs, she suffered a serious setback, and for the rest of her life she remained below average in size. In the breeding pen and the gestation lot she was always the small sow that didn't fit and didn't look right.

The record book, however, showed that it was actually the other sows that lagged behind her. Across 10 litters she had an 11.2-pig weaning average and averaged over three keeper boars per litter.

Field Notes

Perhaps the place to begin a discussion of swine records is in the farrowing house. Here two great opponents meet: the hog farmers who believe they are too busy to put things down on paper, and those who are drowning in a sea of data yet thirsting for more numbers to crunch.

Nailed above every crate gate in our old eight-crate pull-together was one of those little clothespin-type paper clips. Snapped into each one was a 3- by 5-inch index card. Onto that card went the sow's ID number, the boar to which she was bred, her farrowing date, her litter size, their birthweights, and our comments on the sow's performance.

We have seldom in 30-plus years of hog raising operated with more than 20 sows, but even with a sow herd you can number on one hand there can be too much data to keep just in your head. Those index cards might have wound up fly specked, written on in three different colors of ink, and dog-eared, but they were easily at hand for quick notations where the action was really happening.

When the sows were pulled from the farrowing quarters, the notecards were taken to the house and deposited in a file box for future reference. For an investment of less than $2 we were able to build a performance record on our most important asset, the sow herd.

The shirt pocket notebook is a handy way to keep field notes.

Another couple of dollars buys you a little notebook termed a "shirt pocket handbook." Available from purebred swine groups and many mail order farm supply houses, it contains a gestation table and a log page for each litter farrowed. It has spaces to record sire and dam, farrowing date, and more, plus a space to record each pig born by its litter and individual notch.

It thus becomes the place to begin amassing information on

your individual pigs. You can log in such data as teat counts, health treatments given, and comments on each pig's merits. In the pen and on paper the pigs will begin emerging as individuals.

Permanent Records

This shirt pocket reference is often backed up by a desk model into which the data is transferred to become part of your operation's permanent records. Into this desk notebook go the field notes and end-of-the-day observations on everything from pen assignments to ration changes.

Dad paired his shirt pocket notebook with one of those big wall calendars that banks and elevators used to give away at Christmas time. There was a large square around each date with room to write all sorts of things. Over morning coffee Dad would transfer the previous day's notes onto the calendar, and by the end of December he had a yearlong log that could be rolled up tightly with his notebooks, closed with a rubber band, and stored neatly in a drawer for future reference.

Today those notes are just as apt to go on a computer disk, but the point is that a permanent record system is important. As I write this, my permanent record is less than an arm's reach away from my favorite chair. It helps me keep track of the little points that can make enterprise management so very much easier.

For example, while at the 25-sow level we operated for a time with two herd boars. With 25 all-red sows and two all-red boars the stage was certainly set for at least some confusion, and we lost track of the service sire of a couple of litters. Some pretty good purebred pigs had to go into the butcher pen simply because things were left to memory or written down on scraps of paper torn from feed sacks that ended up in the washing machine instead of the record book.

Take the Drudgery Out of Record Keeping

For years I relied upon 3- by 5-inch index cards and pocket notebooks kept in small tin file boxes, along with a desk model herd book. (Some extension agencies offer a version of the desk book at a quite modest cost.) In a matter of a few minutes at the end of each day, field notes and observations got filed away or entered into the permanent log. There is no real drudgery in maintaining such records.

Record-Based Management Decisions

Filing away all of this data in a neat manner until tax time is really just a small part of operating a swine enterprise with good records. You also have to rely upon them to guide you when it comes time to make tough choices.

With 15 to 20 sows, the butcher hog market going south at $1 and more a day, and corn pushing through $3 a bushel, the producer has to face up to some rather hard decisions. How deep to cut? Which animals to cut? Which areas to cut back in? Where to spend money to do the most good? Yes, sometimes the best thing to do during troubling economic seasons is spend some money.

I often make a very strong statement that gets the juices flowing in just about any group of farmers, and I will stand by it here, too: There is not a herd or flock of any livestock species anywhere that would not be made better by simply removing the bottom third of the animals that it contains. Sometimes the "my herd is bigger than your herd" pride gets in the way of good management decisions. But your owning and operating costs would be trimmed dramatically by removing animals on the low rung, and your income would not be trimmed nearly as much you might think. The poor performers always rob from the attention and resources that should be directed to the better performers in any herd or flock.

A Hog's Worth May Be More than Meets the Eye

How to know which animals to remove is the challenge. Just strolling through the herd and choosing strictly on visual appraisal can give disastrous results. The thin sow with the razor-sharp backbone would look like a sure candidate for culling, but she may have just weaned a litter of 10 big pigs and done as well with her previous three litters.

The next sow you encounter may stick in your mind for having weaned nine pigs, but what if three of them weighed far less than the herd average for weaning weights? Next up are two females that have just weaned eight-pig litters, but for one it was her first litter, for the other, her eighth.

Only by having good records on current and past performance can you make the hard decisions. You can't trust every litter farrowed to memory, let alone every pig, and if you rely only upon sight appraisals you might just wind up with a lot of slick-haired sows that farrow a three-pig litter once every 12 months.

Sow herd. While on the subject of breeding herd records, it is probably a good time to go into some detail on the subject of sow herd performance. Nationally, the average sow herd turns, or replaces, 40 percent of its numbers yearly, and many of the commercial gilt suppliers expect a 40 percent culling rate of the females they supply before even the second parity.

As a sow ages, her farrowing times tend to grow longer, and there is an accompanying increase in the number of stillbirths in each litter. You will see the normal wear and tear of aging, too, including some unproductive udder segments. Over time some sows will also succumb to some sort of injury or malady that impairs their reproductive performance.

I will generally give a gilt that farrows a litter of six or seven a second chance (a couple of them have turned into sows that regularly farrow litters of seven). A sow with a small litter will also get a second chance to redeem herself if the problem was obviously due to management or environment. For gilts, a good number for both litter and weaning size is eight; for sows, it is nine. Still, it is best to acknowledge a problem performer in the early going, get her out of the herd, and bring in a younger replacement with better prospects. A few years ago, we rebuilt our Duroc herd and went through nearly a dozen females before finding the one female to build upon.

Average Length of Service

We had one sow that was with us for seven years and 12 litters. But most sows' usefulness lasts three or four years on a small farm, and less in a large commercial operation. Most boars are gone by 30 to 36 months of age, because they have grown so large — often over 500 pounds — and because by then, their daughters are coming into the herd.

Financial Records

Good financial records make it possible not only to determine your production costs but also to pinpoint weak areas within your enterprise. Repeated studies of producer records across the country show that often the only difference between the most profitable producers and those just scraping by is how well they do at that most basic of all management practices — purchase for the swine enterprise.

An old but valued tool for tracking and recording income and expenses is the multiple-column record book. Available for a couple of dollars from most stationery stores, it provides spaces for recording the date of the purchase, what was purchased, from whom, the amount of it, and the cost; additional spaces allow you to break it down by specific category (grain, protein supplement, veterinary, etc.). This data can be recorded in just a few minutes at the end of the day or when bills are paid.

Much of this is information already carried in the farm checkbook. By breaking it down in a columnar fashion, though, you can backtrack through your enterprise in a number of ways that will help you assess its health and success. Records can give a good overview of everything from what it costs you monthly to stay in business to gradual creep in the costs for very specific inputs. On most small farms, the primary expense to produce pork is feed costs, and the largest expense there is for grain. On some of the larger farms and pork factories, interest and energy costs are beginning to rival feed costs.

If costs are beginning to creep up unduly in a specific category, they will jump out at you if you use this type of record keeping. Records can also help you target areas in which cash savings may be possible. At this writing, for example, the elevator east of our hometown is charging 20 cents more per bushel for yellow corn than the elevator west of town. Price columns that don't move when they should denote the need for some comparison shopping for a particular input.

With a few months of such records in the book, you will be able to compute costs to maintain a brood sow, produce a 40-pound feeder pig, or crank out 100 pounds of market hog. Those figures are key to assessing the financial well-being of our venture.

Study Records Monthly

Study your records at least once a month for trends. Too many producers look upon records as something to work at to please the tax man rather than something they can use to improve their enterprise.

Analyzing Recorded Costs

If costs to produce are exceeding available selling prices then your venture is indeed ill-fated. And if all costs to support said enterprise are not directly traceable to it, it may actually be subsidized by another kindred or connected venture. If your hogs, chickens, and cattle are all being fed out of the same reserves of corn and you have no way to ascertain how much of that grain is going where, you indeed run that risk.

There are two sets of costs involved in swine raising: fixed costs and variable expenses. Fixed costs are the costs of owning hogs that have to be paid year in and year out. They include interest and payments on principal, taxes, insurance, building costs, and the like. Variable costs are those that the producer can do something about and include feed, veterinary bills, energy, transportation, maintenance, and supplies.

Smaller-scale producers too often pay too much for basic supplies and services. There is not much to be had in the way of discounts when feed is bought just 200 or 300 pounds at a time (though some feed dealerships do offer 1 to 3 percent discounts for cash purchases). What is interesting to note is that the supposed volume discounts offered to many large-scale producers really do not exist. The industry has been hard pressed to demonstrate any buying advantage for producers even to the 1,000-sow level. Real advantages exist only for producers large enough to totally bypass an entire group of suppliers.

Several years ago I worked in livestock feed sales, and seldom did we have more than a few-dollars-per-ton leeway for dealing with our largest customers over our smallest. Our very best prices came as a result of such things as crop increases and competition among suppliers. The big operators did seem to take greater advantage of such specials when we had them available, however. They had the storage space and perhaps the greater liquidity to move on such bargains when they did occur.

Monitor Variable Prices

The time to take a look at variable prices is not when the increases are coming fast enough and in big enough increments so as to be painfully obvious. Instead, scan your records for emerging problem patterns at least once monthly.

I can recall one summer when large acreages of soft wheat in Missouri and Illinois sprouted problems with wild onions. The wheat harvested from those fields was so laced with the onions that it could only be used in livestock feeds. Some 12 percent complete swine finishing rations actually slipped to well under $90 per ton whether you bought 1 or 100 tons. A place to store it and the money to buy it were all it took to take advantage of such an opportunity.

On feeds that are consumed in small amounts, regardless of operation size, there are few opportunities for volume purchase discounts. Even in the Midwest, some elevators order and stock swine feeds in volumes as small as 500 pounds.

A good long-term set of records can help to point up things like seasonal buying advantages for various inputs. Sometimes, for example, the best time to buy straw for bedding is at the end of winter, when demand is down and sellers are clearing storage space for the new crop of straw coming that summer. However, don't buy into every cliché. At harvest is often the best time to buy a supply of grain, but in years of really short crops the best buys may not come until the following summer.

Recording Sales

Along with inputs, you should keep equally detailed records on your sales. The columnar notebook allows space for recording the date of the sale, type of sale, to whom the sale was made, selling price, and amounts sold by category (butcher stock, feeders, breeding stock, cull animals).

Contrary to what many believe, the advent of high-volume swine producers has not eliminated the traditional seasonal aspects of swine production and their consequences. Feeder pigs still bring the highest prices in March and April, because few producers want the difficult task of farrowing in the cold winter months. Their prices are most depressed in early summer and late fall, following the spring and fall farrowing patterns, that are preferred.

There is a general downturn in red meat prices in November, when all of those spring-farrowed pigs are coming to town and the big holiday meals build on a turkey. June, July, and August often bring the highest butcher hog prices of the year, because the demand is up for barbecue meats while the pork supply is down, because there are fewer cold-weather-farrowed pigs.

Knowing who bought what and how well they paid helps you build reliable markets and know where to target the various segments of your production. Does volume buyer John Smith readily pay market top for 60-pound feeder pigs, but hedge on 40-pounders? Why? Didn't the ABC buying station pay more for cull sows than the XYZ order buyer?

And then there are the kickers in the deck. Are enough sales falling in the month the farm payment is due? Are you dumping cull stock in late fall when all red meat prices trend downward? Are hogs fed in one season of the year taking substantially longer to get to market than those sold in another?

The most interesting reading on the small farm should be the record books. They give the look beneath the hood that is vital in assessing how well that farm is doing and where it is heading.

Ten Crisis Management Steps

In a situation where herd reductions are forced due to plunging prices or drought conditions, there are steps you can take to make an orderly retreat that preserves your herd as a viable unit:

1. Recognize the emerging problem as soon as possible, so that you can take steps before salvage values plummet.
2. Cull older sows first whenever possible. They represent the most dated and what should be the least-productive genetics in the herd. The age of older sows is usually about three and older.
3. Females in their first three parities must be considered the future of the herd: They have the longest productive lives before them.
4. Apply the best remaining feedstuffs where they will do the most good: to females in late gestation, nursing females, and young pigs.
5. Consider alternative feedstuffs, which may be cheaper at times (e.g., replace a portion of the corn in a ration with grain sorghum, or use lysine to replace a portion of the soybean oil meal).
6. Adjust weaning times to best utilize available feed supplies. (When necessary, I've dropped weaning to 28 days.)
7. Buy feedstuffs only as needed.
8. Sell the pigs as feeder stock.
9. Use the depressed prices to upgrade your breeding herd by buying herd replacements on a down market.
10. Pat yourself on the back for managing the situation.

Marketing Hogs

Marketing is the neglected chore on nearly every hog farm. Producers approach it with dread, little knowledge of what drives their markets, and a belief that no matter what they do the market will always be stacked against them.

Many years ago, local radio stations reporting the hog and cattle markets from the National Stockyards in East St. Louis, Illinois — just across the river from St. Louis — would give the names of the sellers at the market that day who had the higher-selling animals. It was quite a feather in the cap down at the coffee shop to have landed the high price of the day and be mentioned on the radio.

At first glance, setting the market-topping price would seem a very valid goal, but it can be one with a painful backlash. Producers who hold out for the very last penny that the market has to offer often find the market buckling rather than following their expectations. A more reasonable goal would be to make all of your sales within the top 10 percent of market prices for the year.

The basic cull stock, butcher hog, and feeder pig markets function as wholesale markets. They are set up to move a lot of volume, and all too often the only premium they have to offer is for maintaining that volume. Buying stations across the Midwest often hold an extra 50 to 75 cents per hundredweight under the counter to draw those numbers when they are needed. And they favor those who can provide the numbers with that premium.

Quality Counts

Still, in recent years the market has begun to at least pay lip service to questions of carcass yield and quality. In the early 1990s there were many rumors of buying stations and order buyers turning away butcher hogs because they just didn't measure up. A lot of small farmers were nipped by this, and many overreacted to the point of discontinuing a basically sound swine enterprise, or backing away from starting one. They had fallen into the trap of thinking that they were just too small to compete anymore, too small to pay for quality inputs, and too small to rise to any challenge. It was a costly mistake — but one with a fairly simple and inexpensive solution.

The answer to the quality challenge is often just one good boar away. During the hard going in the early 1990s I sat ringside at more than one auction where that kind of boar — of good quality but a less-popular breed — sold for as little as $200.

The swine industry wasn't built on $100,000 buildings and $25,000 pickups — it's the hogs, stupid. A valued banker friend once told me that a sure sign of a hog raiser on his way to trouble was a request for a loan for a big, flashy farrowing house. In fact, dollars in any shape or form other than at the bottom line at the end of the year are a pitifully poor way of scoring a venture. A few years ago a Duroc boar made lots of waves with a $35,000 selling price, but just 18 months later I bought one of his sons for only $125.

There are boars out there that can shave off ¼ inch of backfat from their offspring, or add 1 square inch to the average loineye area. Some of them are like pepper on a good steak — you just need to use a dash — but the only way to make hogs truly better is to breed them to better hogs. It matters not a whit where they were born, what they ate out of, or what kind of truck they rode to town on — it is the blood and breeding that tell the tale.

Marketing Appeal

Ten good butcher hogs will draw a good dollar just as often as 100 in an area with competitive market outlets. A small group of butchers can be given added appeal in several ways:

- ◆ Have pigs sorted closely and be uniform in both type and weight. In a small group, a pig out of the loop by even as little as 15 pounds will show up vividly. Information from a national survey conducted by the USDA Animal and Plant Health Inspection Service shows that product uniformity is very important in the pork industry. Packers often give premiums to producers who bring uniform-size pigs of similarly lean quality for slaughter.
- ◆ Be a market watcher. Ten head are easier to load than 100 and can be gotten to market quickly on days with low runs, or when prices take a sudden jump.
- ◆ Don't back away from incentive programs. Too many small producers respond to programs such as grade and yield buying, which pay a premium for quality hogs, with an unfounded fear that their hogs can't cut it on the rail. Actually, on the rail is the fairest place to compete.
- ◆ Don't try to hide a dog in the middle of a bunch of good hogs. Instead of trying to slip a poor hog past a buyer, acknowledge it as a mistake you can and should eat, or at least send along to the sale barn.
- ◆ Know your market people, listen to their wants and needs, and, as much as possible, try to produce the hogs they like to buy.

Note: The truest view of hog quality emerges when they are slaughtered, dressed, and awaiting processing: Their fat cover and lean yield can be measured exactly. The hogs are suspended from the packing house "rail" and the carcasses fully exposed for quality assessment.

Many packers now wait to pay for butcher hogs until they see them on the rail, then offer fairly modest premiums for those that grade and yield better than the average for hogs of the same weight. The greater the red meat yield, the more valuable the hog is to the packer, and the greater the premium payment to the producer.

The poorer-quality pig will generally bring as much as it can at the market of last resort — the sale barn. There are, however, buyers who specialize in odd lots and misfits, and may risk trying to increase such a pig's weight or quality through additional care and feeding.

Study the Market

Doing a better job of marketing often means acquiring more knowledge about how your available market outlets operate. For example, at times we have taken one or two large boars to market after using them in the breeding herd to produce one, two, or three pig crops.

The fate of most such boars is pepperoni and other highly spiced sausage products, but not every market outlet has access to buyers for such specialized production. Often, mature boars are held at a local assembly point, enduring much stress and shrinkage, and then moved along to a regional market. I found one local buying station that bunched such boars for shipment to an East Coast processor and paid as much as $2 to $3 more per hundredweight for them than other available markets. An extra $10 to $12 per head for a couple of animals sold each year may not seem like much, but the first boar this processor bought paid for the phone calls to find which market was paying the most for heavy boars, and that knowledge became a little bit of a perk that I was able to pass on to the buyers of our young boars. In a year or so they would have similar boars to sell and, hopefully, would remember who gave them the tip about this market, and return to us for new boars.

A bit of study of your local scene can reveal the many things that make local markets rattle and hum:

◆ At most markets, highest prices come in the first four days of the work week. Mondays and Tuesdays are generally the strongest ones.

◆ Prices generally trend lower on Fridays, since the buyer usually has to carry the hogs through the weekend before processing them. The same is true of hogs sold on the eve of a holiday.

◆ Some markets have a broader definition of handy weights than others. Some will be more receptive to hogs under 230 pounds or over 260 pounds.

◆ Some markets eagerly seek cull breeding stock, others buy them as a courtesy or a cost of doing business, and others won't handle them at all. The prices each offers reflect its interest in this market segment, which is quite substantial. One area buying station won't buy them, but will ship them to a larger, regional market if space is available on their trucks. One neighbor sent a big boar this way and, due to shrink and fighting stress, suffered nearly a $100 loss in the boar's value in just a few hours.

There are market outlets that do reward loyal patrons and not just with the checks that they write. We once sent 10 thin sows to an area buying station at which the operator directed the trucker to take them to a weekly auction a couple of miles farther down the road. There they brought 12 cents a pound more than the station buyer was authorized to pay — from a local farmer, who bought them to glean his cornfields.

Marketing Trends

Most finished hogs — more than 55 percent of them — are sold directly to the slaughter plant. Only about 11 percent of U.S. hog businesses sell for slaughter through auctions, according to the USDA Animal and Plant Health Inspection Service.

Hogs also travel a way to market, although this varies with the region. More than 80 percent travel 200 miles or less to slaughter, and more than half travel 100 miles or less. They travel farther in the Southeast, but in midwestern and north-central parts of the nation, about one-third are within 50 miles of a slaughter plant.

Special Markets

Much is made of what is now being called "niche marketing," but selling to rather narrow, very specialized markets has been a selling option small farmers have tapped into for generations. Some of them are rather short term, if not singular selling, but others have grown into family businesses with the potential to go on for decades.

Here's an example of a short-term market option. I once had two little Spotted pigs go into a young girl's Easter basket and received a $10-per head premium for them. As an example of a marketing option that persists, I recall a story about a modest-size purebred producer. He began supplying roasting pigs and even doing the roasting for parties and celebrations as a means of marketing hogs that were not of the quality needed for breeding stock. This market grew to a level of two or three such events each week and soon replaced purebreds as his primary swine venture.

Selling show pigs. A rapidly growing specialty market in many parts of the country now is the raising and sale of show pigs. These are pigs in the 35- to 60-pound weight range and of such quality as to be considered good candidates for the show ring. With a good many families moving to a few acres in the country and becoming part-time farmers, a growing number of youngsters are seeking project animals but do not have the time or space to give to a traditional sow and litter project. At some county fairs the market hog show classes have totally replaced the breeding stock shows.

Some producers are even organizing into marketing groups to sponsor once- or twice-a-year auctions of such pigs. As many as 10 producers may come together, consign 10 head of pigs or so each, hire an auctioneer, select a sale site, and pay for advertising collectively. A number of pig producers may even join with an equal number of club lamb producers to sponsor such a sale. Some purebred producers are even forgoing the traditional fall sale to offer up some of their best spring pigs in a March or April show pig auction.

The show pig market is seasonal, with pigs of the right age for the summer and fall fairs being the most in demand. This generally means January-, February-, and March-farrowed pigs for sale in March, April, and May.

Pigs sold for show are the cream of the crop, so early on sales might be small. Among the most active in this market are the small and midsize producers with some previous experience in fitting and showing hogs.

Barrow pigs need to have been castrated early and fully healed. While not exactly in a fitted state, shoats are presented in a condition over and above that normally seen in barn-run feeder pigs. They are absolutely free from parasites and are kept well housed and bedded to maintain quality haircoats. They are the big pigs in the crop, display exceptional muscling, and are free from structural faults. They may be purebred or F_1 animals that just missed the cut as breeding animals.

Show pigs are not from thrown-together matings. They reveal a lot of black (Hampshire, Berkshire, or Black Poland) and red (Duroc) breeding, to take advantage of the growth and muscling strengths of those breeds. Their numbers are limited at an auction, to maintain quality and to prevent a price break from occurring in such a public setting.

To further increase demand, some producer groups even sponsor their own jackpotted market hog shows. They schedule a show with a date that will catch the pigs they have to sell, offer a top prize (sometimes the only prize) of between $250 and $1,000, and open the show only to pigs sold by group members. This show might be scheduled for a week or two ahead of another event such as the county fair to give the youthful patrons a bit of added experience. Other groups may offer a $50 or $100 cash award and trophy to any pigs from their offering that top various area shows and fairs.

Poland Breed Association

This Poland China boar has good length of body and depth of side. He shows good testicle development and a trim penile sheath. He's got a really good ham, too, and he's a boar that should do a good job of siring show pigs.

Show Pig Producers: On the Cutting Edge

Show pig sellers are also offering a bit of expertise. Pigs selected for sale should be of currently popular types; the producer makes him- or herself available to the young buyers to answer questions about fitting and showing; and the producer attends the various shows to lend support to their buyers.

Show pig producers are a bit on the cutting edge. They tend to be knowledgeable about bloodlines, are breeding current types, and are familiar with the show ring scene.

Pork Sales

A market we are beginning to plumb with some of our Willow Valley hogs is the pork trade. A local abattoir produces a very lean, whole-hog pork sausage from our hogs that we sell by the pound at our local farmers' market. It is a USDA-inspected and -approved facility, and it processes, wraps, and fast-freezes the product for us.

We're no threat to Bob Evans or Jimmy Dean, but we do wind up with a premium product that we're able to sell for $2 a pound or better. We have the pork ground 80 percent lean to 20 percent fat (quite a bit leaner than most store-bought varieties) and only mildly seasoned. Spices can be added to sausage right up to the time it is eaten, but removing them to suit someone's taste is quite another matter.

The sausage is frozen by the slaughterhouse into 1-pound sticks, and we pick it up and transport it to the market in simple ice chests. These markets might be the local farmers' market or doorstep sales. I know of one farmer who sells sausage products from a freezer carried in the back of his pickup and plugged in whenever the truck is parked at a market site. At least in Missouri, you do not need a special license to do this as long as the meat comes from a USDA-regulated facility. We use the local processor's packing label, but our own label with details about the lean content and additive-free nature of our sausage would, no doubt, help sales. Alas, such custom-made packaging is very expensive and has to be bought in considerable volume.

There are some of us who share a fairly common sentiment that family farmers should be freer to directly process and sell their own production. They can't, however, because of red tape. This figure may sound a bit high, but I would say a farmer should be allowed to process and direct market the pork output of up to 500 head of hogs per year. That would be just 1,000 cured hams and bacons to sell each year — hardly a threat to even the smallest of supermarkets.

In a direct-selling situation, you as producer have even more of a responsibility for quality control than if you were answerable to a whole phalanx of tax-and-license-fee-financed inspectors. You have to look at the end consumers of your product and know that your success comes from their satisfaction and that alone. Two quality cured hams will now often bring as much or more than an entire butcher hog on the hoof — but family farmers are now largely cut off from this market, which was once their traditional domain, through a series of nitpicking rules that do not always prove effective in guarding public health or food quality.

Pork sausage is just one pork product that has proved to be a good venture for small hog farmers.

Custom-Feeding Hogs

Farmers preferring a less-involved approach to direct-marketing butcher hogs offer whole or half carcasses delivered on the hoof to a processor mutually agreed upon by the buyer and seller. I have seen these hogs advertised as "fed to order" or "custom pork product." The buyer can request a specific slaughter weight, additive-free rations, range raising, and the like, and the producer prices the hog accordingly. A substantial deposit is in order before undertaking to custom-feed a market hog. It is certainly a practice in keeping with the trend toward community-supported market gardening and subscription fruit growing.

A listing under a "Good Things to Eat" heading in the want ads of a nearby metropolitan paper may be among the best places to locate such buyers until word of mouth (the very best advertising) can build. Simple flyers pinned up in places such as health food stores may also draw potential buyers.

Marketing Tools

I have used the weekly newspapers in the counties adjoining ours to offer boars, gilts, and feeder pigs with good results. Calls to the swap shop programs on local radio stations have also paid off with sales for me. I put up flyers at local sale barns, too; and you could try passing out business cards at busy places such as elevators or auctions when you hear someone expressing an interest in acquiring hogs.

We maintain a list of all of our past buyers, including their addresses and phone numbers. When a set of young boars is ready to sell, I can check that list for those who should be due for a new set of boars. Then I send them a postcard with a listing of what I have to sell.

One midsize breeder I often work with mails out his own newsletter. It features brief items and pictures of his buyers and lists the animals he currently has for sale. When his sales slow, he offers discounts on the purchase of two or more boars or a sale on older boars; once he even included a $25 off coupon good on the purchase of one boar.

A most imaginative selling tool he uses makes many buyers feel more at ease with selecting boars from him while also steering them to his higher-priced animals. The pigs he deems best in a group receive purple ear tags, and the second level gets blue tags. Buyers see this as a way of reinforcing their choices.

The colors were chosen for a purpose. The very best hog in a show receives the purple ribbon and it comes out of the top group, the blue-ribbon winners.

On another farm, if 10 gilts are bought at the regular price, an 11th is available for just $100. Years ago, one legendary breeder would actually give a free boar with every 20 gilts purchased.

We too give discounts to buyers — $10 to $25 off per head for purchases of two or more head, and a similar discount for repeat customers. This generally works well, but at 8- to 16-month intervals I have a set of brothers who take 10 to 12 boars at a time. They expect a bit of preferential treatment and they get it, but too much of a discount and it's like giving them a boar free.

More Marketing Ideas

It makes very little sense to work hard at all of the other phases of swine production only to falter and fail at the last and most important task, marketing. Imagination and a dedication to quality will stand the small producer in far better stead than simply playing the numbers game.

Many are uncomfortable with the practice, but you must promote your pork enterprise at every opportunity. Pork may be promoted nationally as "the other white meat," but the only one who will promote your hogs and your pork is you. Toward that end, keep in mind that:

◆ In friends, neighbors, family, and business associates you have a ready-made list of potential buyers.

◆ Opportunities to promote come in odd ways. I know one pork producer who brings a new and novel pork dish to every potluck dinner his family attends.

◆ You need to let people know where you are. Get your name on the mailbox and get a sign out in the road pasture. If they don't know you're there, they can't buy what you have to sell.

◆ Group efforts can pay off, whether you're establishing a farmers' market, starting a sale, or just getting enough hogs together to fill a big truck.

◆ Good publicity helps your business grow. Create your own press releases or seek help from those with whom you interact, such as breed association staff. We once bought the top-selling boar at a breeder's auction and got a nice mention in the breed association magazine, a couple of state farm magazines, and a couple of local newspapers.

◆ It helps to be seen in the right places. We attend Pork Day activities in the next county, take in nearby sales, attend hog shows, work with 4-H youngsters, and buy from our fellow small producers.

The Hog Industry: Today and Tomorrow

Earlier, I pointed up the risk to small farmers of thinking they are disadvantaged or unable to compete because of the small size of their ventures. Quite simply, mindset has been either the undoing of or the key to success for more smallholders than anything else by far.

It is far easier to fund and establish 10 different small farm ventures netting $2,000 each yearly than to create just one enterprise with a capacity for $20,000 in net earnings. On the small farm, at least a couple of those modest ventures can be swine based. We raise breeding stock, sell feeder pigs, and market pork sausage, and all from quite a small group of sows that fit into a number of other ventures on our small acreage.

We putter along with a sow herd that falls through everybody's net, a lot of 20-year-old equipment (pile it all up, light a match, and for $1,500 I'm back in business tomorrow), and own nary a pickup or tractor. Still, we produce a quality product that breeds on for our customers. We're putting money in the bank, and we're content with what we're doing. And never underestimate the importance of that last point.

Growth: At What Price?

Much, too much, is made of growth in production agriculture. As I watch those about me pushing to add ever-higher numbers and build ever-bigger herds, I'm reminded of two things. The first is a line my grandmother often repeated: "No bird ever flies so high but what it doesn't have to light." The other gets even more to the point — growth for growth's sake is nothing more than the philosophy of the cancer cell.

There is absolutely nothing wrong with staying small, as long as you are committed to staying good. At times you may feel that you're striking out on a lonely and diffident course, but such isn't really the case at all.

The average sow herd in the United States still numbers less than 40 head. Pigs are on farms from Hawaii to Delaware in numbers great and small. Just one sow producing 18 pigs in a year translates into over 1½ tons of pork sausage. If you have a market for that much sausage at $2 per pound, you have gross yearly sales in the $6,000 range.

You do, that is, if your pork is of good quality, an approved processor is available, and you are comfortable with direct-selling and extracting that high level of production from a sow. At this writing, the small-scale pork producer must function as something of a maverick — a re-pioneer.

Before I get accused of coining a totally new word, what I mean is that smaller pork producers are in a position to lead an entire industry back to the point where the animals are humanely reared, the environment well tended, the consumer better served, and all producers respected as worthwhile, contributing members of rural society.

Public support of production agriculture has begun to markedly slip. The quality of its output is being questioned and challenged as never before. The only farming sectors still in total favor with the consuming public and whose output garners a willingly given premium from that public are the organic growers and those producers who can be clearly identified as family farmers.

Corporate Competition

Conversely, from the farmer's perspective there is often doubt whether the small producer can survive, let alone continue to compete. Right now it is hard to ignore the big corporate producers — they are the 800-pound gorillas — but their warts are beginning to show. Close-packed swine populations are more disease prone, they are creating great seas of wastes, they sell entirely in a wholesale market, they have fixed operating costs that are often as high as the variable costs on some small farms, they have odor problems that make them the classic NIMBY enterprise, and consumers are at the point of organizing to resist their existence.

There is some strong evidence that the big operations are seeing the writing on the wall and may be planning to get out of the pork business sometime in the first 10 to 20 years of the next century. Many of the big buildings that have gone up have done so with a 10- to 15-year cost recovery plan. These operations are often just single arms of multinational corporations that can shed them as easily as some lizards drop their tails.

If you don't think this is the case, just look at the broiler industry: Incubators could be turned off, hatching eggs dumped, workers cut loose, and the farms would still chug along just fine with all sorts of other income. On the other hand, few contracts for swine finishing extend for more than a year, and one of the largest corporate swine operations in the Midwest applied for bankruptcy protection a scant five years into its run.

Animal rights issues, environmental measures, and consumer concerns can and may all grow to a point where the pressure they will bring to bear on the large operations will be just too great. Some small farmers are already tapping into this resistance by pursuing niche markets for additive-free pork, pork raised outside, and pork with real pork taste. The niches of today are the only alternatives for tomorrow.

The Personal Touch

Studies have shown that the way a person handles hogs is important. Pigs that receive pleasant and empathetic handling, for instance, are easier to manage, grow faster, and perform better than those that received unpleasant handling. It's the kind of care that the small farm operator can provide.

The Family Hog Farm

I honestly believe that nothing will be sacrificed by a return to production entirely from family farms. Note these favorable points:

- Small herds mean big litters of good pigs. Small-scale producers have the time to see that everything gets done right the first time, and to know and manage each animal in the herd as an individual.
- They produce hogs as a part of a diversified plan of production. The big corporations offset startup costs and downturns in the price cycle with income from other sources; small farmers can do the same with income from other enterprises on the farm.
- Small diversified farms have a level of environmental integrity that the big, specialized outfits can never aspire to. The wastes from 10 sows go back to the fields with a manure spreader. The wastes from 1,000 sows go into a lagoon that may develop problems with the next big rain.
- Small size in no way precludes quality. Pick up a 60-year-old ag text and you will find page after page of references to "ton litters," 10-pig litter averages, and hogs hitting a 220-pound market weight at a bit over five months of age — and all of this performance was fueled with open-pollinated corn, skim milk, and tankage. If small farmers did this once, they can do it again.
- Small farmers are discerning users of technology and not slaves to the methodology of the moment. They don't have to have the latest bit of techno-puffery to help them cope with the health, ventilation, or waste problem of the month.
- They can and do work with quality inputs. Time and again, breed-building boars and sows have come from small farms, and the family farmers are now the ones guarding the gene pool that the seedstock companies draw from to create their syntho-composites.

An Amish Example

The ability of the small-scale swine producer to survive is perhaps best epitomized by the Amish example. West of the Mississippi, Amish and Mennonite farmers have embraced pork production nearly as often as they have dairying in the East. And they do it with a clear-eyed vision of what they are about.

Over the years I have worked with Amish farmers as a field representative for the Hubbard Milling Company, as well as supplying them with purebred breeding stock. The Amish farmers I've come to know operate at levels of from 10 to 60 sows and produce feeder pigs, butcher hogs, and crossbred breeding animals. They were among the first producers in our area to regularly use F_1 breeding animals and follow crossbreeding rotations.

The Amish farmers I've worked with and continue to supply are very receptive to new ideas that are practical for them and will help them to obtain "optimum production." They read the hog magazines, study type trends, give new products a fair trial, invest in quality breeding stock, use simple, easily maintained housing and facilities, and use pork production to best and fully utilize available family labor. *Optimum production* means seeking a fair and reasonable return on investment rather than attempting to totally maximize production in faint hopes that returns will eventually surpass costs. In other words, by *optimum* I mean aiming for a fair and livable return, rather than attempting to extract every cent of return possible.

I've seen $500 boars go to Amish farms with but 10 sow herds, have known Amish producers to contact breeders 100 miles from home and order boars sight unseen to get genetics they needed to move ahead (they did their genetic homework first), and delivered to them some of the newest in feedstuffs and animal health products. Within the use of just a couple of hundred pounds of a new feedstuff or a test on one set of pigs, these farmers can discern if a product or method will work for them.

Defining Optimum Production

To illustrate the optimum production concept, let's take apart the old industry chestnut that every sow has to wean at least six pigs per litter just for you to cover all costs and break even. It's still not a bad rule, but in an expensive facility with a totally controlled environment that figure is now more like 7.5 pigs per litter needed to break even.

Interestingly, there are some folks out here weaning six pigs per litter and making money at it. They're farrowing in the brush, have a couple of $25 fence chargers bought at auction, are farrowing in 20-year-old huts, follow a more seasonal farrowing pattern, use a well-thought-out plan of crossbreeding, and market their production rather than just trying to get something sold whenever a note payment comes due.

The swine industry is now wrestling with the question of just how many pigs a sow can produce in a year. There are producers in Europe bumping 26 pigs per sow per year, and some believe that 30 or even more is doable. Very early weaning, multistage nurseries, and sows that must be supported with red hot diets and rigidly controlled environments make this doable, yes, but at a very great cost. There are sow lines now that can no longer be penned outside during the changing seasons, nursery buildings run at a steady 90°F year-round — no matter how high the cost of propane gets — and pigs being weaned at 10 pounds rather than the usual weaning weight of 25 pounds at 35 days, or 40 pounds at 56 days. How many producers are there who can realistically afford to be this good?

In nearly every operation there is room for improvement, but there is also that point at which further increases in production will not carry the greater costs to produce. The time and resources needed to wring every dollar possible from a venture are simply not always justifiable on economic grounds. And let's be realistic here — life is just too short to spend the whole of it down in the hog lots rooting out every last dollar to be had there.

Archaeological data keeps pushing back the date when hogs were first domesticated. Certainly they have moved across this earth in the company of humans for a long, long time. Suckling pig laid on the tables of kings and a cured ham swung over the saddle of a westward wanderer: Pigs and pork and people have formed quite an alliance.

Ensuring Your Future

The future for hog producers is just as great today as it has always been. There will be a place in it for pork producers of all sizes — if they will accept but a few simple tenets of the successful swine herd:

- ◆ You have to like 'em even when they don't smell like money
- ◆ Start small, but start well
- ◆ Grow as you learn and only to a level at which you are comfortable
- ◆ Maintain quality animals that perform well and are a pleasure to own
- ◆ Produce them with a very specific plan of action, from breeding pen to market

Eleven

DAY-TO-DAY LIFE WITH HOGS

A COUPLE OF YEARS BACK, I was invited to speak about my hog-raising experiences at a seminar for beginning small farmers and those seeking new ventures to add to existing small farm enterprise mixes. On the morning program, I was sandwiched between an exotic species broker and an ostrich farmer.

Not wishing to distract interest from the speaker following me, I excused myself following a short question-and-answer period. I was greatly surprised, however, at the number of folks who followed me out of the meeting room and continued their questions in a nearby trade show aisle. Hogs seem to at once intrigue, frighten, and inspire those folks.

One man was eager to make a start with hogs, but his wife was concerned about odors with a tenacity that would do a pit bull proud. Another was having difficulty finding certain breeds close to his home farm. A third man was just bogged down in unfavorable "hawg" imagery. I steered the group to a nearby display of live breeding hogs. A few minutes of hands-on experience allayed many of their fears. No one was bitten or succumbed to noxious fumes. Breeders were found who would deliver breeding animals bought from them, or at least cooperate with delivery to central points. And the hogs endured our intrusion into their space with all of the dignity inherent in their race.

Honestly, I know of no substitute for hands-on experience whatever the agricultural venture you are considering. A few head of hogs are themselves best teachers of what they need. They will give you the title "hog farmer" and, in time, will give you the savvy and experience to merit it.

That is the gist of this chapter: being a good hog raiser, a good small-farm-based hog man or woman. Working with hogs can also sometimes be a humbling experience — as this true story will attest.

Sage Advice

A group of us supposed "veteran" producers were sitting together at a purebred hog sale watching three ring workers struggle to cut a couple of gilts out of a larger group of 250- to 300-pound gilts in the sale ring. One of us ventured that they needed more cutting gates and hurdles: another said more men were needed; and a third ventured that more gates and men both were needed. A truly wise old gentleman sitting behind us and high up in the bleachers put us all in our place when he correctly surmised that what was really needed was to replace at least two of those halfway useful ring workers with one good stock dog.

Movin' Hogs the Easy Way

Actually, the most difficult and time-consuming task on a great many farms with hogs is getting them from point A to point B while maintaining a modicum of dignity and your religion. I have actually seen films of a farrowing house rigged up with a length of 12-inch-diameter PVC pipe through one outside wall at a downward angle, to speed up pig handling by allowing weanling pigs to be slid through the pipe and into a waiting truck or trailer. In a 180° turn from this attempt at a pig elevator, we had an elderly neighbor who would place a bushel basket over a sow's head and back her up and down chutes and into farrowing stalls with comparative ease. Keep 'em in the dark and keep 'em moving was his philosophy.

When I pull shoats from weaning quarters to move them 300 or 400 yards to growing pens, I put them into a plastic trash can on wheels. Two or three young shoats will travel quite comfortably in this unconventional, but quite inexpensive and simple-to-use "trailer." It also will hold a whole litter of very young pigs while I give them health treatments, or a few boar pigs while I sort through a litter for keepers and hold the rest for castration.

For larger hogs, I often use a "farm boy" hog trailer made from one 16-foot-long hog panel and a 5- to 6-foot-long segment cut from another. I bow the longer panel into a horse-shoe shape, then securely wire the shorter panel across the open side. This simple unit can be used as a small holding pen, a sorting aid inside a larger pen, and, with the help of a second person, a tool for moving even a large sow or boar over short distances.

It can also be tipped over a hog, or the hog can be driven into it and the short panel wired shut behind. The lightweight unit can then be dragged by the narrow front end by one person. A second can walk along behind, urging the hog forward and quickly pushing down on a side or end if the animal balks or attempts to root beneath it. Quite often, the

A plastic trash can on wheels can be used to move two or three shoats quite comfortably for a short distance.

gentle bumping from behind with the short panel is all that's needed to keep the hog moving to a new pen or pasture.

As noted earlier, the best way to keep a hog moving is to block the way behind it. Also, as much as possible, allow the hog to move along at its own pace and in the company of other animals to help it remain confident and at ease. It is often far easier to drive two or three hogs back off a trailer than to drive just one hog onto it.

Beware of Light

When you're moving hogs, it helps to think ahead and even to think a bit like a hog. Changing light patterns, intensities, or shadows in their line of sight can be most disturbing to animals being driven from one place to another. Especially difficult can be the task of driving a hog from a darker barn environment into bright sunlight, and vice-versa.

In areas where hogs will be regularly worked, or along frequently used lanes of travel, many producers line the fencing or enclosure with ply-wood or sheet tin to completely block out shifting light patterns and other outside distractions. Hogs will also move up solid-floored loading chutes more easily than those with slatted floors.

Managing Information

Perhaps even more challenging than getting hogs moved from pen to pen with a fair semblance of order is the task of keeping current on the hog scene.

The information age has come to the countryside with a vengeance; the material available often seems to flow over small farms in great waves. At last count, we received five monthly magazines that dealt solely with hogs, three others with regular inserts that targeted hogs, and at least a half a dozen general-interest farm magazines with extensive hog coverage. This doesn't even begin to include the various catalogs, newsletters, extension bulletins, and books that regularly find their way to our mailbox.

I awake to market reports on the radio, eat dinner (lunch to you folks outside the Midwest) with market commentary in the background on radio or TV, and finish the day with the TV weatherman helping me to select my work attire for the next day. Phyllis, my supportive wife, knows that I have at last grudgingly acknowledged the arrival of winter when I heed the weatherman's warnings and don what she calls my "long lingerie."

In talks with those contemplating a hog venture of some sort, I find that they are too often put off by this floodtide flow of ever-changing and sometimes seemingly contradictory information. It has reached the point where a single case of Peruvian Higgedly-Piggedly disease in Taiwan can receive multimedia coverage in the United States within a matter of hours, and be in the print media in detail within a very few days. Fortunately, there is no such thing as Peruvian Higgedly-Piggedly disease, and information management must now become just one more aspect of good farm management and operation.

I've yet to find a hog that can read and I trust that if I do find one so gifted he will be wise enough to take everything he reads with a grain of salt. In 30 years at the hog game, I've seen a great many trends, products, and practices crash and burn within months of having been named "can't miss" in the farm press. Some have even come around again a time or two. Two that rise quickly to mind are the "flat-muscling" and "late-maturing" trends of about 15 years ago.

"Flat-muscled" hogs looked to most folks like the hogs that were once called "cat hammed" and "meatless wonders." "Late-maturing" hogs were also largely "slow-growing" hogs. Of course change is sometimes good. For example, the new role of the Berkshire breed in producing the highly sought-after "black pork" for the Asian trade looks like something that will benefit all sectors of the pork industry. Still, type changes and new production methodology often seem to be advocated only when certain sections within the industry need something new and/or pricier to sell.

Count on Other Producers

I want to see what I read about being tested in the field before I buy into it. Fortunately, swine producers as a group are very open and sharing people. Many magazines actually include the addresses and phone numbers of producers mentioned in their articles. I have called complete strangers to ask them about their first-hand experiences with a certain new product or bloodline. It is also a favor I have done for others like me, who are trying to take a sounding on the latest on the swine scene. Time your calls for between 8 and 10 P.M. on weekdays (don't forget the differences between time zones), keep calls short and limited to just a couple of questions, and include a self-addressed, stamped envelope with all letters of inquiry.

I also know a number of long-time friends and fellow swine producers whose input and opinions I rely upon greatly. When the phone rings at 8 or 10 P.M. on a weekday, it's generally one or another of our group plugging in to talk over what's happening locally, regionally, or nationally, or seeking help in working through a problem of one sort or another. We try not to overextend our welcome through this practice, but these kinds of sounding boards have proven vital in working through what will and won't work for our operation.

Staying routinely in touch with other hog farmers is one of the best ways to stay up to date and solve problems.

Better Information Management

Here are five steps to help you better manage information.

Step 1. Invest a reasonable amount of regularly scheduled time in information acquisition and management. I have often heard it said that the really good farmers spend an average of two hours daily reading and otherwise processing the information currently available to them. It is a figure I certainly believe to be correct. Tucked into the magazine rack next to your favorite easy chair should be a large envelope with pen or pencil, highlighter, and scissors. Into the envelope can go articles or even whole issues you wish to keep for future reference. It's easier to take clippings on first reading than to stack magazines away thinking you'll get back to them at a later date. Dad and I would even make marginal notes and pass magazines back and forth.

Step 2. Do your own field-testing on a small scale. Make things prove themselves on your farm, which is like no other on earth. A great many farmers have invested tens of thousands of dollars in new technology only to see it all fall out of favor in far less time than it took to come into vogue.

Step 3. Plug in to your fellow producers for their input and experiences.

Step 4. Go see it and go see it again before you buy. If a practice or product or animal catches your eye, study it in detail, and then bring it onto the farm in the smallest, simplest, and easiest-to-manage way that you can. Dad once traveled across three counties looking at different types of pig creep feeders, came back home, sorted through our scrap pile, and built one himself. We used it for several years before replacing it with one of the factory models he liked, which we were then able to buy used for just $25.

Step 5. The best money you can spend is the money that buys you hard and solid information. A $25 book can often save you thousands of dollars in costly mistakes. It should also be pointed out that publications and periodicals pertinent to your particular field of endeavor are generally deductible expenses on your Schedule F federal tax form.

Do It Right the First Time

I seem to recall an old razor blade commercial that said something to the effect that to be sharp you needed to look sharp. On the farm with hogs looking sharp means doing things right the first time.

It has often amazed me how that lone, short length of 2 inch by 4 inch board leaned up in the corner of a farm building is soon joined by a pitchfork with a broken tine, an old 5-gallon bucket, a busted tank sprayer, and a push broom with the handle flamboyantly wrapped in silvery duct tape. We all need to be more mindful of the three Ps for good farmstead keeping: *Pick* it up as soon as it drops or falls, *put* it back as soon as you're done with it, and keep it *painted* or otherwise maintained.

Good tools and facilities do indeed seem to improve the quality of a farm's output. They can also greatly improve the efficiency of the operation, because there is no substitute for the right tool for the job. They create a pride of workmanship that can be almost tangible. A clean and neat farmstead also inspires confidence in the potential buyers who may visit you.

I will concede that, all too often, my own shop and buildings do harbor more than their fair share of "boars' nests," but I was taught better. First and foremost is the safety factor for both the producer and his or her porcine

We all need to be more mindful of the three Ps for good farmstead keeping: Pick it up as soon as it drops or falls, put it back as soon as you're done with it, and keep it painted or otherwise maintained.

charges. Lose a needed handtool or cause a tear in an expensive bag of pig starter and you can swear now and laugh later; cause an injury to yourself, a family member, or a valuable breeding animal, however, and the consequences can be quite far-reaching. A barrow I was once heavily counting on to take to a local hog show injured its eye on baling wire carelessly left dangling in the pen after tying off a gate. Within minutes, the animal went into shock and was dead.

Step on a nail and you may be out of action for several painful days. Let that same mishap befall your one and only herd boar and you are out of business.

There are times when my management style is as loose as a bucket of ashes, so when mistakes and accidents happen, there is no one to blame but myself — the creature in our hog pens who is supposed to have the biggest brain. Every farm has those catchall spaces where old gloves and slip-joint pliers seem to migrate of their own accord, but too often these are also the spaces where producers first go to look for things as dear and delicate as livestock health supplies.

Did you ever wonder why some swine health care products are packaged in brown glass bottles and others are in clear glass? The ones in the brown bottles are light sensitive. Leaving them on that handy shelf at the barn or the dash of the pickup will certainly shorten their useful life, and may even render them totally useless.

There is an old saying that in the deep woods where the bears and wolves reside, it is still the mosquitoes that drive back most of the intruders. The small things can indeed be the undoing of farm operations of all sizes. Do the little things in a timely fashion and they won't grow into the crushing tasks that may eventually overturn your entire operation.

A Well-Run Farm

One of the best-run small farms I ever visited not only had a refrigerator in an outbuilding for the correct storage of veterinary pharmaceuticals, but on that fridge door was a list of what was inside, when it was bought, when it would expire, when it was last used, and how much of it remained. A veterinarian of my acquaintance measures the skills and stockmanship of his clients by how clean they keep the tops of their bottles of injectable livestock drugs.

Take Time to Scratch Their Ears

Several times each day, there is nothing more important that swine producers can do than simply go out and look at and listen to their hogs. Early in the morning and late in the day (around sunrise and sunset) are excellent times to take a really hard look at the animals in your care. At these times, give your animals your full attention rather than attempting to note things simply in passing. It is at these times of the day that soundness and respiratory problems should be most noticeable. They are also good times to note the early listlessness that so often foretells developing health problems. Those animals that are slow off their beds or that distance themselves from others in the group should receive a much closer inspection and perhaps have their rectal temperatures taken, to help form an early diagnosis of sorts.

This is also a feel good thing for most farmers with livestock. Seeing the hogs well fed and content at the end of the day can take the edge off all those aches and pains you've been feeling. The animals are your hopes and dreams for the future given form and substance, and you have every right to feel content and rewarded as you stand viewing them.

One of the most important things that swine producers can do is simply go out and look at and listen to their hogs.

Size Up Your Hogs

You need to spend time really looking at your hogs, comparing them with that mental standard or ideal that all good producers carry around in their heads. Duly note those animals that don't measure up to that standard, and mark them for elimination from the herd. An old proverb says that nothing fattens the beast like the eye of its owner; we all have to learn to look past what we want our hogs to be and see them as they really are.

Pull out that pocket herd book or draw on your memory and start matching up pigs to their parents. Are old Sally's pigs growing like you thought they would, or is it time to start thinking about how many pounds of pepperoni on the hoof she represents? Are those feeders that cost so much to buy muscling up the way you anticipated, or is it time to look for a new pig supplier?

Goals

I have never met a hog farmer who was completely caught up with chores. We all have something more to do, or something more we want to incorporate to add value to our swine ventures. And we've all had days when we labored mightily just to get to a point where we felt it was safe to leave the hogs for naught but a warm meal and a few hours of badly needed sleep.

This wanting to make things better should be an incentive and an element of pleasant anticipation within your swine venture. Don't beat yourself up emotionally for not achieving everything here and now. The story is often told of the farmer who met the visiting county agent at his gate and told him not to even bother to come in, because he already knew how to farm better than he could afford to.

I believe in the concept of stringing together small, doable goals to create an ever-improving operation. Too many farmers work only for that one big day when, somehow, weather, markets, crops, and genetics will all come together to make them a great success in one fell swoop. Buy into that one and you will burn out or die with a broken heart.

In the early going we worked to add one or two good, home-raised gilts every few months, or to add a little better boar than the one before it at intervals of 18 months or so. I raised a lot of hogs in our old horse barn before I bought my first new hog house, and I raised purebred Durocs for nearly two years before adding a second sow to that venture. Our first boar with a recorded pedigree was bought "used" from a neighbor who was holding back gilts and no longer needed him.

Work Slow and Steady

Slow and steady on the farm are ways to both learn and grow. Weaning an 8-pig average should make you want to try for an 8.5-pig average next time, not to invest in a $60,000 farrowing house to hold dozens or more sows to wean 8 or fewer pigs per litter. A goal should be something attainable in a reasonable amount of time — not something that places the whole future of your venture in peril should it fail to materialize.

Have an Eye for Details

The real strengths that a small-scale producer has to bring to swine production are time and attention to even the smallest of details. We do not put off until tomorrow because we know that that kind of thinking will take away from profits today.

The classic example of this strength manifesting itself is in the average litter size to be found on the nation's small farms with sows. These are the natural home for truly big litters, because where litters are few in number, every pig is precious, and efforts to save them are unstinting.

To increase output from the farrowing unit, many producers follow a practice called cross-fostering. They place nursing pigs with sows and in litters where they will have the greatest potential to thrive. This is also one more reason for farrowing females in closely timed groups.

For the sake of an example, let's take a hypothetical producer with 12 sows divided to farrow into two groups of 6 sows each. He breeds each group of sows to farrow within a very few days of each other.

He enters his farrowing house early one morning to work with the six sows there, and finds that all six had farrowed within 120 hours of each other.

Three have good litters of 8 to 10 evenly sized pigs. One is an older, heavy milking sow with just 7 large pigs, but with a good track record as a pig raiser; another is a second-parity sow with 11 rather unevenly sized pigs; and the last is a gilt with 6 big pigs and 2 smaller ones. By making a few number adjustments among the latter three litters, the producer can produce more even pigs at weaning, increase pig survivability, get optimum production from all sows, and get these six sows bred back as a uniform group for the next farrowing.

Here are 10 tips to successful cross-fostering:

1. Don't switch pigs receiving colostrum to sows no longer producing colostrum, and vice-versa.
2. Try to switch only those pigs born no more than 36 hours apart.
3. Contrary to what many might think, the best pigs to move from litter to litter are not the small ones but the largest and most vigorous pigs in the litter.
4. The bigger, more vigorous pigs should go to a sow that can milk well, to maintain their current level of performance.
5. Small pigs should be grouped with a sow that has a proven track record as a pig raiser and as long a potential lactation period before her as possible.
6. Get litters as even as possible while moving as few pigs as possible.
7. Handle the little pigs very gently, to keep them from squealing or showing other signs of distress when they are placed in a new pen and with a foster dam.
8. Try to place little pigs on the new sow while she is nursing.
9. Monitor the newly restructured litters for several hours following relocation of the pigs. Any not making the transition can be returned to their dams.
10. In cross-fostering, many believe that scent can be a factor in getting a sow to accept her new family. Toward this end, a lot of little pigs have been liberally doused with those Christmas talcs and aftershaves that most of have stored away over the years. These lend some interesting new odors to the farrowing quarters. Still I believe that the normal tumbles and turns of a litter of little pigs soon has them all smelling the same anyway.

Don't Pack 'Em Too Tight

I pointed up earlier the need to have a place for every tool and always return it there. The same holds true for the hogs themselves.

I believe that on even the smallest of farms there can be room for several livestock and poultry ventures of modest scale — but not all in the same place. Many years ago, there was a book that had as its central theme the maintaining of a great many livestock varieties in a single building or barn.

The appeal of such a practice is readily evident, but how successful this single-site loading would be in the real world is quite another matter. We have hog houses that have been used — at different times — to house everything from a sow and litter to growing hogs to ewes with lambs to baby chickens, but never in any combination.

In a shared barn, little pigs just don't last very long beneath the stamping, kicking feet of cattle and horses. And small calves and large hogs don't make good penmates or neighbors.

Poultry ranging free through a barn can spoil bedding and feedstuffs, damage equipment with their droppings, and unsettle some hogs with their activities. Also, over the years I have heard many reports of sows that developed a strong taste for chicken tartare. Chickens past one year of age can also harbor the avian tuberculosis germ, which may be transmitted to the hogs.

Everything in one barn sure sounds like a good idea, but it was a practice not even pursued in pioneer times, when livestock housing was at an even greater premium than it is now. They knew that on free range, hogs, cattle, and other livestock species would fare better than when crowded into a single facility.

Buying at Auction

Throughout this book, I have made reference to buying quality used hog equipment. One of the best places to find such equipment is at a farm auction. It is a fast-paced atmosphere in which to attempt to make purchases, however.

Over the years, I have assisted my in-laws in conducting a number of farm sales they booked as professional auctioneers. Through observation and from working the ring at a good many farm sales, I have picked up a few practices that help to ease the chore of buying at auction:

- ◆ Carefully study the sale bills, to determine if the offering contains items that will fit your needs.
- ◆ Try to visit the sale site a day or two before the actual auction. You can then take longer to examine in detail the items in which you are interested. The owner will also have more time to answer your questions about how the items were used and maintained. When you're

buying used livestock equipment and housing, it's especially important to determine if the equipment has been recently exposed to any disease-causing organisms.

◆ Arrive early on auction day, and dress appropriately. Don't wear the clothing and footwear you usually use for chores. Hog owners and their vehicles from all over may be in attendance, so the potential for contact with any number of potentially harmful disease organisms is great — many of these can be transported on boots, clothing, and the undercarriages of vehicles. Also, dress for the weather. Often sales have to go on whether the snow flies or the rain falls, and those dressed to stay and stay comfortable to the very end may reap some real bargains.

◆ Don't start bidding without a fixed top bid already in mind. Don't go above that bid no matter how much the auctioneer and his crew wheedle and cajole. An old rule of thumb that serves many well is to spend no more than half of their new price for used farm goods.

◆ Don't hesitate to bid first if the offering is something you really want. People are reluctant to start bidding for fear they'll look too eager, but I have seen far too many items fall to the first bidder to hold back on something I want. Bidding first may be the only chance you have to put some kind of definition on the bidding range.

◆ In many instances, it's a good idea to offer a bid somewhat below the one the auctioneer is seeking. If the auctioneer has a bid of $10 in hand on a nose snare, is asking for $15, and the bidding has slowed down, it may be time to offer a bid of $12.50 or even $11. A time-honored bidding gesture is the slashing of one index finger across the other to indicate a bid increase of just half of the one the auctioneer is requesting.

◆ Step back when you've reached your bidding limit. This will shake the auctioneer off and let it be known that you're done. Don't feel self-conscious about standing by your limit. Most veteran auction goers will respect you for your savvy approach and well-thought-out bidding.

It's As Is!

Remember, the merchandise at an auction usually is sold "as is." In fact, the terms of sale at most farm auctions are "as is, where is." If you buy it, then go to pick it up and the bottom falls off, you'll learn to look more closely next time. Any expressed warranties are strictly between the buyer and the seller; the auctioneer is the agent for the sale and nothing more. If the seller wishes to hold a reserve bid or floor price on an item, this must be announced as the item goes on the block or it has to be sold to the highest bidder. Following the sale you will have from 7 to 30 days to move the goods from the premises. Short days should be announced by the auctioneer as a part of the terms of sale. The auctioneer should also be able to assist you in locating truckers to haul your purchases if needed.

When the hammer falls and the auctioneer announces an item sold to you, it is your baby 100 percent. Should it be stolen, run over in the parking lot, or sat on by an elephant, it's your loss to take. Sadly, you are more and more apt to encounter theft in even the most rural areas these days, and the smaller items at auction seem to be especially vulnerable to theft in all of the confusion.

Hog Farm Fashion

Fashion might seem like a funny thing to talk about in a book about hogs, and I will grant that, rather than fashion plates, most hog farmers are likely to be fashion saucers. Still, I have already noted the importance of proper dress for attending an auction, and, for all of your tasks, the right clothing and footwear can and will do much to add to your comfort and safety.

For example, there is far more than simple mud and muck in spring and fall hog lots. These lots are liberally laced with urine and manure, which can take a toll on even the best shoes and boots. Further, when your feet are wet, chafed, or otherwise discomforted, the rest of you is out of sorts, too. If you regularly wear leather footwear, make sure to buy work shoes that have been treated to be manure and urine resistant. There are also a number of leather shoe treatments that will help maintain their suppleness and resistance.

There have been years — 1993 and 1994 come to mind — when it seemed that I wore knee boots every day except the Fourth of July. The slip-on styles are easy to get on and off, and they do a pretty good job of keeping your lower legs dry, along with your feet. You can spend a lot for a pair of knee boots, but due to the ever-present risk of tears and punctures while doing farm tasks, I prefer to wear less-expensive boot models. It seems to be the Klober law of rubber boot ownership that the more expensive the boots are, the more likely I am to punch a hole in them.

When I first became involved with purebred hogs in the late 1960s and early 1970s, many of us affected a sort of British herdsman look. We wore green knee boots, a coverall with full sleeves topped by a khaki or duck coat or vest, and, on our heads, a short-billed, brown duck Jones cap. The back of that cap could be snapped down to keep rain from going down the back of the neck.

It was warm, it was durable, and, by golly, we looked like hog farmers. It was a style of dress that endured until the "gimme" advertising caps appeared on the scene and we opted for a denim-rich, flannel-heavy "urban farm boy" look. We still hung onto our rubber boots, and Dad often teased that I should quit worrying about supposedly free caps, go where the feed was the best buy, and use part of the money I saved to buy my own headwear. It was also good advice.

Denim and flannel still dominate today's choices in work clothes, but producers in temperature-regulated buildings are opting more and more for short-

Author Kelly Klober in his current-day work clothes.

T. L. Gettings /RSI

sleeved coveralls (I can't bring myself to call them jumpsuits). It's sort of like Star Fleet designed a uniform for the manure-spreading set.

All kidding aside, good work clothing that fits well and is made of durable fabric is one of those small but very important things that ease your labors. Very loose-fitting clothing can be a safety risk, because it can be more easily grabbed by spinning shafts and moving gears and pull you into them. The traction and grip provided by quality footwear also add greatly to your safety.

The Crease in the Cap

Midwestern farmers do add personal touches to all of those multicolored, mesh-backed caps given away by feed and farm supply companies. You can sometimes get a clue as to where they hail from in the Corn/Hog Belt by the way they crease the bill of their caps. The sharp crease in the center of the bill seems to be the Missouri look of the moment; across the big river in Illinois, they appear to favor two creases, on the outer edges of the bill, giving it a squared-up appearance.

Thus far, hog farmers have avoided those dinner-plate-size belt buckles and reptile-skin boots, but I would like to share a short joke that every winter seems to make the rounds of all those supper meetings the feed companies use to introduce their new products. Do you know why hog farmers don't wear athletic shoes? Because feed companies don't give them away.

Hog People

There is no finer group of people on the land than this nation's family farmers with sow herds or related ventures. Invariably, I have found them to be considerate and sharing people, willing to give freely of their time and experiences.

They know what it is to live high on the hog and low on her hocks — yet I have never met a single producer who was hardened or had a real root-hog-or-die attitude toward life. When you enter their ranks, you will find yourself a part of a real fellowship based upon shared experiences and a commonality of goals.

You will have to earn the mud on your boots and the dings in your pickup, but you are part of an industry that was spawned with Columbus's second voyage to the New World. You may cuss 'em up the loading chute one day and cry over them the next when pigs in a promising litter are lost despite your best efforts. But then, that's farming.

They don't sing songs about hog farmers the way they do about cowboys, but those of us with mud on our jeans and hogs on our farms can take a great deal of satisfaction in knowing that in this protein-starved world, we're producing one of the leanest, most healthful, and most nutritious of all foods.

I like hogs. I've also made a lot of payments with their help, and met a lot of good folks through them. I still see in them a promising future for small farmers willing to tend them in a thoughtful and caring way.

Gene Hebert/American Livestock Breeds Conservancy

People with the time and talent to be good hog raisers also seem to have the best of human virtues and, indeed, they make the best of neighbors.

APPENDIX A

REPRODUCTIVE INFORMATION

In this appendix I provide some basic reproductive information, to help you determine times of estrus, gestation, and farrowing.

Reproductive Facts

- The duration of estrus is 1 to 5 days, and the average is 2 to 3 days.
- The regularity of estrus is 18 to 24 days, with the average being 21 days.
- The first occurrence of estrus after farrowing will take between 1 and 8 weeks, but can then be expected to occur every 17 to 24 days until the sow is bred. Two litters per sow per year is a reasonable goal, and some producers are achieving 5 to 6 litters in a 24-month period with weaning of the pigs at 5 weeks of age or less.
- Gestation will take between 110 to 116 days, with the average being 113 to 114 days.
- Puberty for gilts can occur between 120 and 235 days of age, with the average being 200 days of age. There are some differences between breeds as to the onset of puberty in gilts.

Gestation Table

This information will help you determine when farrowing will occur.

Breeding Date	Farrowing Date	Breeding Date	Farrowing Date
Jan 1	Apr 25	Jul 1	Oct 23
Jan 15	May 9	Jul 15	Nov 6
Jan 30	May 24	Jul 30	Nov 21
Feb 1	May 26	Aug 1	Nov 23
Feb 15	Jun 9	Aug 15	Dec 7
Feb 28	Jun 22	Aug 30	Dec 22
Mar 1	Jun 23	Sep 1	Dec 24
Mar 15	Jul 7	Sep 15	Jan 7
Mar 30	Jul 22	Sep 30	Jan 22
Apr 1	Jul 24	Oct 1	Jan 23
Apr 15	Aug 7	Oct 15	Feb 6
Apr 30	Aug 22	Oct 30	Feb 21
May 1	Aug 23	Nov 1	Feb 23
May 15	Sep 6	Nov 15	Mar 9
May 30	Sep 21	Nov 30	Mar 14
Jun 1	Sep 23	Dec 1	Mar 25
Jun 15	Oct 7	Dec 15	Apr 8
Jun 30	Oct 22	Dec 30	Apr 23

Pedigree Form

Breeder _____

Address _____

Telephone _____

Sire _____

Ear Notch _____

Color _____

Wt. _____

Winnings _____

_____

Sire _____

Ear Notch _____

Color _____

Wt. _____

Winnings _____

_____

Sire _____

Ear Notch _____

Color _____

Wt. _____

Winnings _____

_____

Dam _____

Ear Notch _____

Color _____

Wt. _____

Winnings _____

_____

Sire _____

Ear Notch _____

Color _____

Wt. _____

Winnings _____

_____

Dam _____

Ear Notch _____

Color _____

Wt. _____

Winnings _____

Breed _____

Born _____ Sex ____

Name _____

Ear Notch _____

Color _____

Wt. _____

Winnings _____

_____

Sire _____

Ear Notch _____

Color _____

Wt. _____

Winnings _____

_____

Sire _____

Ear Notch _____

Color _____

Wt. _____

Winnings _____

_____

Dam _____

Ear Notch _____

Color _____

Wt. _____

Winnings _____

_____

Dam _____

Ear Notch _____

Color _____

Wt. _____

Winnings _____

_____

Sire _____

Ear Notch _____

Color _____

Wt. _____

Winnings _____

_____

Date _____

Sold to _____

Address _____

_____

Dam _____

Ear Notch _____

Color _____

Wt. _____

Winnings _____

_____

Dam _____

Ear Notch _____

Color _____

Wt. _____

Winnings _____

I certify that this pedigree is correct to the best of my knowledge and belief

_____

You can copy this blank pedigree and use it for your hogs. Record the ear notch of your pigs to avoid inbreeding by positively linking the pig to its sire and dam.

Monthly Management Chart

Post in a location that will be convienient to record the facts about your hog raising. Monthly sheets can be totaled to determine an "annual report" of your hog production.

	Expenses					
Date	**Animals Purchased**		**Pounds of Feed**	**Cost of Feed**	**Other Costs: Supplies, etc.**	
	No.	**Cost**	**Used**		**Item**	**Cost**
Total						

	Income	
	(List income from sale of equipment, breeding fees, etc.)	
Date	**Item**	**Amount Received**
	Total	

Monthly Management Chart continued

Breeding Record

Date Bred	Name or Numbers		Date Due to Farrow	Date Farrowed	Number of Live Young Born
	Boar	Sow			
				Total	

Show Participation

Date	Name and Place of Show	Placing or Award Received	Entry Fees	Value of Premiums Won
		Total		

RESOURCES

The importance of keeping current on swine trends and staying in touch with others in the industry has been emphasized throughout this book. Here are some resources to help you toward that end.

Periodicals

Acreage Advisor
15400 North 56th Street
Lincoln, NE 68514
800-516-5359

Alternative Agriculture News
Henry A. Wallace Institute for
  Alternative Agriculture
9200 Edmonston Road, Suite 117
Greenbelt, MD 20770
301-441-8777

BackHome
P.O. Box 70
Hendersonville, NC 28793
800-992-2546

Center for Rural Affairs
P.O. Box 406
Walthill, NE 68067
402-846-5428
(A newsletter surveying events
  affecting rural America)

Countryside
W11564 Highway 64
Withee, WI 54498
715-785-7979

Hogs Today
Center Square West
1500 Market Street
Philadelphia, PA 19102
215-557-8937
Fax: 215-568-3989
http://www.hogstoday.com/

Missouri Ruralist
1007 North College Avenue
Columbia, MO 65201
573-895-5445

Modern Small Farm Newsletters
99 Hawks Ridge Drive
Troy, MO 63379

National Hog Farmer
7900 International Drive, Suite 300
Minneapolis, MN 55425
612-851-4710
http://www.homefarm.com/

Pork
P.O. Box 2939
Shawnee Mission, KS 66201
913-438-8700
Toll-free, 800-255-5113
Fax: 913-438-0695

Progressive Farmer
Box 830069
Birmingham, AL 35283
205-877-6494
Toll-free, 800-292-2340
Fax: 205-877-6450

Rare Breeds Journal
P.O. Box 66
Crawford, NE 69339
308-665-1431

Seedstock Edge
P.O. Box 420585
Palm Coast, FL 32142
Small Farm Today
3903 West Ridge Trail Road
Clark, MO 65243
800-633-2535
Fax: 573-687-3148

Today's Farmer
201 Ray Young Drive
Columbia, MO 65201
573-876-5205

Supply Houses
Familiarizing yourself with a variety
of supply houses will help you find
better prices.

Cuprem, Inc.
P.O. Box 147
Kenesaw, NE 68956
800-228-4253
(swine health products and odor
   control products)

Eco-Shelter, Inc.
Route 2, Box 59
Friend, NE 68359

FarmTek
1440 Field of Dreams Way
Dyersville, IA 52040
319-875-2288
Toll-free, 800-327-6835

GLMS Industries
East Highway 50
Peabody, KS 66866
316-983-2136
(pig shelters)

Innovators
10801 Highway 36 North
Brenham, TX 77833

Kane Manufacturing Co., Inc.
P.O. Box 774
Des Moines, IA 50303
515-262-3001

Kent Feeds
1600 Oregon Street
Muscatine, IA 52761
319-264-4211
(feed)

Luco Manufacturing
P.O. Box 385
Strong City, KS 66869
316-273-6723

McBee Agri Supply, Inc.
16151 Old Highway 63N
Sturgeon, MO 65284
800-568-4918
(electric fence supplies)
Midwest Livestock Systems, Inc.
3600 N. 6th Street
Beatrice, NE 68310
888-342-5657

Modern Farm
1825 Big Horn Avenue
P.O. Box 1420
Cody, WY 82414
800-443-4934

Nasco
901 Janesville Avenue
Fort Atkinson, WI 53538
414-563-2446

Port-Λ-Hut
14 Peterson Drive
Storm Lake, IA 50588
712-732-2546

WXICOF
914 Riske Lane
Wentzville, MO 63385
314-828-5100

Internet Sites

For those of you who are computer savvy, the Web will provide a vast array of information. Use one of your search engines to search for "swine" and you'll be amazed at all the resources you can access. Here are some of the Web sites that are available.

Animal and Plant Health Inspection Service (part of the USDA)
http://www.aphis.usda.gov
Coop Extension Sites
http://www.reeusda.gov/new/csrees.htm

Homesteading and Small Farm Resource
http://www.homestead.org
This Web site has everything from information for producers to information on the nutritional value of pork, recipes, and special sections for kids.

National Pork Producers Council
http://www.nppc.org

Pork '97
www.porkmag.com

United States Department of Agriculture (USDA)
http://www.usda.gov

Breed Associations and Registries

American Berkshire Association
P.O. Box 2436
West Lafayette, IN 47906
765-497-3618

American Yorkshire Club
P.O. Box 2417
West Lafayette, IN 47906
765-463-3593

Chester White Swine Record Association
P.O. Box 9758
Peoria, IL 61612
309-691-0151

Hampshire Swine Registry
P.O. Box 2417
West Lafayette, IN 47906
765-497-4628

Landrace Association, Inc.
P.O. Box 2340
West Lafayette, IN 47906
765-497-3718

National Hereford Hog Record Association
Route 1 Box 37
Flandreau, SD 57028
605-997-2116

National Spotted Swine Record
P.O. Box 9758
Peoria, IL 61612
309-693-1804

Poland China Record Association
P.O. Box 9758
Peoria, IL 61612
309-691-6301

United Duroc Swine Registry
P.O. Box 2417
West Lafayette, IN 47906
765-497-4628

Miscellaneous Addresses
Here are the names and addresses of organizations that may come in handy or be another good source of information for you. Registries or clubs for the leading swine breeds also appear below; others can be located through the National Swine Registry.

American Association of Swine
    Practitioners
902 1st Avenue
Perry, IA 50220-1703
515-465-5255
Fax: 515-465-3832
http://www.aasp.org

The American Livestock Breeds
    Conservancy
Box 477
Pittsboro, NC 27312
919-542-5704
http://web.css.orst.edu/
    Organizations/Livestock_animal/A
    LBC.html

Appropriate Technology Transfer for
    Rural Areas
P.O. Box 3657
Fayetteville, AR 72702

Food Alternatives With
    Relationship Marketing
P.O. Box 186
Wills, VA 24380
540-789-7778

Livestock Nutrition Laboratory
    Services
P.O. 1655
Columbia, MO 65205
888-257-LNLS
(feed analysis)

Missouri Alternative Center
University of Missouri
628 Clark Hall
Columbia, MO 65211
800-433-3704

National Pork Producers Council
P.O. Box 10383
Des Moines, IA 50306
515-223-2600
http://www.nppc.org

National Swine Registry
P.O. Box 2417
1769 U.S. 52 West
West Lafayette, IN 47906

SPF Swine Genetics
1840 North 48th Street
Lincoln, NE 68504
800-541-0481

Cooperative Extension Service

For more information about hogs and programs in your area, write or call the Cooperative Extension Service in your state. This program is affiliated with each of the Nation's land-grant universities and the U.S. Department of Agriculture in Washington, DC, and can provide information on a wide range of topics.

Alabama
Auburn University
109 Duncan Hall
Auburn, AL 36849
334-844-4444

Alaska
University of Alaska
P.O. Box 756140
Fairbanks, AK 99775
907-474-7214

Arizona
4210 North Campbell
Tucson, AZ 85719
520-626-5161

Arkansas
University of Arkansas
Box 4966
Pine Bluff, AR 71611
870-543-8530

California
University of California, Davis
1441 Research Park Drive
University Services Bldg. #110
Davis, CA 95616

Colorado
Colorado State University
Fort Collins, CO 80523
970-491-6281

Connecticut
University of Connecticut
1376 Storrs Road
Storrs, CT 06269

Delaware
University of Delaware
Cooperative Extension Service
Townsend Hall
Newark, DE 19716
302-831-2504

Florida
University of Florida
Extension Service Department
2800 NE 39th Avenue
Gainsville, FL 32609
352-955-2402

Georgia
University of Georgia
Associate Dean for Extension
Conner Hall
Athens, GA 30602
706-542-3824

Hawaii
University of Hawaii
Honolulu, HI 96822

Idaho
University of Idaho
Cooperative Extension Service
Moscow, ID 83844
208-885-6639

Illinois
University of Illinois
Office of Information, Technology,
  and Communication Services
528 Bevier Hall
905 South Goodwin Avenue
Urbana, IL 61801
217-333-6095

Indiana
Purdue University
Cooperative Extension Service
1140 Agriculture Administration
  Building
West Lafayette, IN 47907
765-494-8489

Iowa
Iowa State University
110 Curtis Hall
Ames, IA 50011
515-294-4576

Kansas
Kansas State University
Animal Science & Industries
Webber Hall, Room 218
Manhattan, KS 66506
785-532-1251

Kentucky
Livestock Disease Diagnostic Center
1429 Newton Pike
Lexington, KY 40511
606-253-0571

Louisiana
Louisiana State University
Cooperative Extension Service
P.O. Box 25100
Baton Rouge, LA 70894
504-388-4141

Maine
University of Maine
Libby Hall
Orono, ME 04469
800-287-0274

Maryland
University of Maryland
1916 Maryland Highway, Suite A
Mountain Lake Park, MD 21550
301-334-6960

Massachusetts
University of Massachusetts
Extension Service
212C Stockbridge Hall
Amherst, MA 01003
413-545-4800

Michigan
Michigan State University
Cooperative Extension Service
108 Agricultural Hall
East Lansing, MI 48824
517-355-2308

Minnesota
University of Minnesota
Extension Service
240 Coffey Hall
1420 Eckles Avenue
St. Paul, MN 55108
612-624-1222

Mississippi
Mississippi State University
Extension Service
Box 9610
Mississippi State, MS 39762
601-325-3036

Missouri
University of Missouri
W 234 Veterinary Medicine
   Building.
Columbia, MO 65211
573-882-7848

Montana
Montana State University
P.O. Box 172560
Bozeman, MT 59717
406-994-4371

Nebraska
University of Nebraska
211 Agriculture Hall
Lincoln, NE 68583
402-472-2966

Nevada
Nevada Cooperative Extension
   Administration Office
2345 Red Rock Street, Suite 330
Las Vegas, NV 89102
702-251-7531

New Hampshire
University of New Hampshire
59 College Road
Taylor Hall
Durham, NH 03824
603-862-1520

New Jersey
Rutgers State University
542 George Street
New Brunswick, NY 08903
908-932-1766

New Mexico
New Mexico State University
Cooperative Extension Service
P.O. Box 30003 Dept. 3AE
Las Cruces, NM 88003

New York
Cornell University
Cooperative Extension Service
276 Roberts Hall
Ithaca, NY 14853
607-255-2237

North Carolina
North Carolina State University
Cooperative Extension Service
Box 7602
Raleigh, NC 27695
919-515-2811

North Dakota
North Dakota State University
Cooperative Extension Service
P.O. Box 5655
University Station
Fargo, ND 58105
701-231-7881

Ohio
Ohio State University
Extension Service Department
2120 Fyffe Road
Columbus, OH 43210
614-292-6181

Oklahoma
Oklahoma State University
Cooperative Extension
139 Agriculture Hall
Stillwater, OK 74078
405-744-5398

Oregon
Oregon State University
Extension Administration
102 Ballard Ext. Hall
Corvallis, OR 97331
541-737-2713

Pennsylvania
Cooperative Extension Office
Agriculture Administration
  Building, Room 217
University Park, PA 16802
814-863-3438

Puerto Rico
University of Puerto Rico
Box 21120
Rio Piedras, PR 00928
787-832-4040

Rhode Island
University of Rhode Island
Cooperative Extension Office
3 East Alumni Avenue
Kingston, RI 02881
401-874-2929

South Carolina
Clemson University
Cooperative Extension Service
103 Barre Hall
Clemson, SC 29634
864-656-3382

South Dakota
South Dakota State University
Extension Service
Agriculture Hall, Room 154
Box 2207 D
Brookings, SD 57007
605-688-4792

Tennessee
University of Tennessee
P.O. Box 1071
Knoxville, TN 37901
423-974-7114

Texas
Texas A & M University
104 Administration Building
College Station, TX 77843
409-845-7800

Utah
Utah State University
Cooperative Extension Service
4900 University Boulevard
Logan, UT 84322
801-797-2200

Vermont
University of Vermont
Extension Service
601 Main Street
Burlington, VT 05401
802-656-2990

Virginia
Virginia Polytechnical and State
  University
Extension Administration
105B Hetchshon Hall
Blacksburg, VA 24061
540-231-6000

Washington
Washington State University
Cooperative Extension Service
Pullman, WA 99164
509-335-2933

West Virginia
University of West Virginia
Morgantown, WV 26506
304-293-3701

Wisconsin
University of Wisconsin
Extension Service
Madison, WI 53706
304-262-3980

Wyoming
University of Wyoming
Cooperative Extension Service
Laramie, WY 82071
307-766-5124

GLOSSARY

ADDITIVES: Additions to a ration that is blended to be nutritionally complete. They can include vitamins, minerals, probiotics, medicants, and flavorings. Their purpose first is to promote growth or fully balance a ration, second to prevent health problems from occurring, and third to maintain economically acceptable performance in the presence of disease. Additives are generally mixed into the ration at a very low rate — often between 1 and 5 pounds per ton of a complete feed.

AMINO ACIDS: These are the nitrogenous compounds that are so often referred to as the building blocks of protein. In an effort to better fine-tune rations, many feed companies now seek amino acid balances in their rations rather than formulating feeds on the older crude protein basis.

ANTHELMINTIC: An agent that gets rid of intestinal worms.

BABY PIG: This term refers to the young pig in its first 14 to 21 days of life, when it is still dependent upon a liquid diet, has yet to develop any natural resistance, and cannot generate sufficient body heat for survival.

BARROW: A castrated male. He is the meat animal that is the basis of the pork industry, although gilts do produce leaner carcasses.

BIN: A round or square container made of wood or metal and used to store feed, grain, and feed components. It should provide protection against moisture, birds, and rodents.

BLOOM: A visual indication of good health, growth rate, and feed consumption. Haircoat will be bright and smooth; the body will have a plump, full appearance; and the pigs will be bright eyed and alert. A pig with a head appearing coarse or too large for its body is most likely stunted for its age.

BLOWN APART: This term indicates good internal body dimension throughout the animal's entire length, which is associated with hardiness and the ability to grow well.

BOAR: An intact male that is being used for breeding or being held for that intent. The term covers intact males of all ages.

BODY CAPACITY: This term refers to the internal dimensions of the hog. A deep side, a deep, wide chest, and length of side are all deemed important to good health, growth, and feed utilization.

BONE: This term is used to describe skeletal dimension as exhibited in leg diameter, size of forearm and jaw, and width in the head (which is believed to translate into width throughout the entire body). Bone is important for durability and is the frame upon which muscle — red meat — is to be hung.

BULK: This term is often used to describe feed sold by the ton or hundredweight rather than in the sack. A ton of feed to be delivered to a bin or feeder is said to be in the bulk. *Bulk* may also be used to mean fiber added to a ration for improved digestibility purposes.

BUTCHER: This refers to a hog being readied for or sold on the slaughter market. A butcher hog is the same as a market hog; it weighs from 220 to 260 pounds and usually is five to seven months in age.

CARCASS: The dressed body of a hog. At this time the hog can be most accurately evaluated for health and meat type, because this is when the ratio of fat to lean and the percentage of lean cuts (ham, loin, shoulder, and butt — the cuts of greatest value) that the animal produced can be determined.

CHARGER: The device that controls the electric current in the fence line.

CHILLING: The cooling — with adverse affects — of a pig or hog by rain, wind, damp bedding, drafts, or mud.

CHITTERLINGS: The small intestines of a pig.

CLEAN: In one instance, this term refers to a hog with visual indications of little fat cover: A hog with a trim jowl, for instance, might be described as clean headed. It also refers to ground or facilities that have had no contact with hogs for an extended period; six months is a minimum for ground to be considered clean, because most disease and parasite cycles should have been disrupted within that time frame.

COLOSTRUM: The first milk after farrowing, through which the sow is able to impart some of her natural immunities to the nursing pigs. It is produced for the first 72 hours following farrowing, and, without receiving it, the pig has almost no chance of survival. Cow colostrum has been used as a substitute, but results are still far from the same.

COVER: The fat that the hog lays on beneath the skin as it approaches market weight. Some fat cover is essential for the market, but it is the most expensive gain to put on a hog, and excessive cover will be discounted against.

CREEP FEED: The early feeds, high in sugar and milk proteins, that are offered to a pig while it is still nursing. These feeds are very complex in their formulation and are most often processed into very small pellets or crumble, to make them easier for the pigs to consume. Walk-in feeders that protect such feed from the sows are called creep feeders.

Crossbreeding: The mating of hogs of different breeds to capitalize on the strengths of various breeds.

Crude Protein: This term collectively refers to all the nitrogenous compounds in feedstuffs.

Culling: The removal of animals from a herd, usually due to poor performance or illness.

Days: This is a shortening of the term *days to market* and refers to the time from the day of birth that takes an animal to reach 220, 230, or 240 pounds. This can be determined anytime the animal begins to approach 200 pounds, because rate of gain can then be determined and future gains projected. When this is done, the result is referred to as *adjusted days. Actual days* refers to what the animal weighed on an exact day in its life (most often between the 150th and 180th day of life).

Drove: A herd or group of pigs.

Drylot: A large lot used to maintain hogs. Drylots are dirt surfaced but, due to foot traffic and numbers present, free from vegetative growth. A mature hog should have a minimum of 150 square feet of space in a drylot — the more the better.

Electrolytes: Mineral salts that increase the body's uptake of water and other liquids. They are used to counter problems with dehydration due to fever, scours, or stress.

Embryo: A pig in its earliest stages of growth within the sow's uterus.

Estrus: The period when the female will accept the male and conceive.

Eyeball: To visually appraise an animal.

F_1: The first-generation offspring of a cross between animals from two separate pure breeds. These are the animals with the greatest boost of hybrid vigor.

FARROWING: The act of giving birth.

FARROW-TO-FINISH: Taking pigs from birth to slaughter.

FEED EFFICIENCY: The amount of feed the animal consumes to produce 1 pound of gain. This figure is most often obtained as a pen average.

FEEDER PIG: A young pig — most often between 40 and 70 pounds — produced by one farmer and sold to another for feeding out to market weight.

FINISH: A term used to describe good fat cover and bloom on a hog ready for market.

FINISHING: Feeding out a hog for slaughter, whether it's for your family table or the slaughter market.

FITTING: The process of readying an animal for exhibition.

FOOTING: A surface that will allow the hogs to walk and mate safely and securely. It needs to be roughened and even cleated where a hog is expected to climb. Too-smooth or -wet surfaces can result in foot and leg injuries and splay-legged hogs and pigs. To reduce this problem, hogs can be broken to dung and urinate in one corner of a pen: Wet that corner and transfer wastes there when you introduce pigs to the pen.

FRAME: The height and width of the skeletal system. A good frame gives the hog a large, massive appearance of width, substance, and durability.

FREEMARTIN: A congenitally defective female that will breed repeatedly but not conceive.

FULL FEEDING: Allowing the hog to consume all of the feed it desires daily. This will generally be 2 to 3 percent of the animal's liveweight.

FUTURES: A marketing arrangement whereby the producer sells a set number of hogs for future delivery on a specific date to obtain a guaranteed price for them. This can enable a producer to lock in a profit, but the numbers involved shut out most small producers. A great many speculators are involved in futures markets. I'm like Will Rogers in thinking that in order to sell a hog, you should've raised that hog.

GENETIC POTENTIAL: The inward genetic makeup of an animal, which allows it to improve the performance of its offspring above and beyond the current herd average.

GENOTYPE: The genetic makeup of an animal. It's what the animal can do for you in areas like growth rate and carcass type.

GESTATION: The period of pregnancy, between 110 and 116 days, with the average being 113 to 114. A lot of old-timers express it as "3 months, 3 weeks, and 3 days."

GILT: A female that has yet to bear young. The term *second-litter gilt* is used to describe a young sow carrying her second litter, but this is little more than a marketing gimmick.

GRADING: The sorting of feeder pigs or butcher hogs as to their quality. This is done by government employees independent of the marketing concerns. Grading is an effort to reward producers with a higher premium for hogs and pigs of higher quality and greater worth.

GRIND AND MIX: A process whereby whole grains are ground and mixed with supplements to form a complete ration with bite-after-bite consistency.

GRIPPING TONGS: A hog-restraining device that applies pressure to the back of the neck.

HAND BREEDING: Direct producer supervision of matings without the boar being allowed to run with the sow herd. The desired sow and boar are placed together and, after mating, returned to separate pens.

HERNIA: Also called a rupture, a portion of an organ or body structure has broken through the wall that normally contains it.

HETEROSIS: Hybrid vigor, or the increased hardiness and capacity for growth exhibited by the offspring of two different purebred strains.

HOG: A swine over 120 pounds in weight.

HOGGING DOWN: Turning hogs into a field to harvest standing grain or harvest wastes.

HOT WIRE: Another term for electric fencing, and well deserved — as anyone who has ever brushed against one can attest.

HOVERS: Three-sided boxlike enclosures that fit in and over a portion of the pig bunk. They are accessible only to the pigs through small "pop" holes and help keep little pigs warm. They fit over a heating pad, or an electric heat lamp is placed over a circular hole cut in the top of the hover.

HURDLE: A short, solid gate used when handling and herding hogs.

INBREEDING: The mating together of closely related individuals to concentrate certain traits. As my father-in-law says, "It can bring out the good and any of the bad, if it's there."

INDEXING: A score for a particular performance trait expressed as a number over or under the herd average. The average is often given a value of 100; better performers will score above 100, below-average performers below it.

LACTATION: The period when the female is producing milk and nursing pigs. Producers now allow pigs to nurse from three to the traditional eight weeks. The shorter the lactation, the less draining it is on the sow, although three- to four-week weanings require special facilities for the weaned pigs and may affect the sow's recycling for breeding.

LIBIDO: The sexual drive that causes the male to seek out females, mount them correctly, and breed.

LIMIT FEEDING: The restricting of the amount of energy feedstuffs the animal receives daily. This is most often done to keep brood sows from becoming excessively fat and having farrowing difficulties. It can be done with growing hogs when feed efficiency and carcass trim can be improved by it, but days on feed will be extended.

LINE: A specific bloodline tracing back to a single individual or family of great merit.

LINE BREEDING: Repeat breeding to a single individual (parent to offspring), to concentrate that individual's genetics in a family.

LITTER: The offspring of a single farrowing. The record litter size is well in excess of 30; the national average is a fraction over 7, and has been for a lot of years. Most producers hold that they need at least six pigs per litter to cover expenses, and eight pigs per litter for gilts, nine for sows are reasonable, reachable goals.

LOINEYE: The area, in square inches, of lean meat contained within the loin. This is the most important cut economically — it is the pork chop — and the larger the loin eye, the more it is worth to the consumer.

MARKET HOG: The same as butcher hog. It weighs from 220 to 260 pounds and is usually five to seven months in age when it goes to market.

MEAL: Finely ground feedstuffs or complete feeds. When ground too fine, some feedstuffs can present problems with palatability and ulcers.

MEAT TYPE: Animals demonstrating trimness and good muscling patterns.

MEDICATED EARLY WEANING (MEW):   Similar to segregated early weaning. Pigs are weaned early and separated from pigs of other ages by moving them to an isolated facility to prevent the spread of disease, but sows also are immunized before farrowing.

MIDDLE OF THE ROAD: This is a term of primary concern to the seed-stock producer. It refers to an animal of good type, breed character, and performance, but not of extreme type or a departure from industry thinking. It is a very balanced animal, a rather complete package. The producer of this type is referred to as a propagator rather than a breeder by those who are riding or leading type swings.

MUMMIES: Partially absorbed embryos that are sometimes seen with the placenta or at the time of an abortion. These are embryos that have died in late term, otherwise they would have been totally reabsorbed by the sow's system.

MUSCLING: Visual indications of the hog's muscle pattern. It can be seen in ham and shoulder shape, the length that muscles are carried down the legs, and freedom of motion on top of the shoulders.

MYCOTOXINS: The poisons produced by mold growth on grain and other feedstuffs. The younger the animal is, the more severely it is affected. Loss of baby pigs and abortions are the most common problems that result from mycotoxins.

NATURAL IMMUNITY: The immunity that an animal or herd builds up over a period of time when exposed to disease-causing organisms. Sometimes it is transferred by survival of the fittest, sometimes via the blood during gestation, and sometimes during lactation.

NATURAL VENTILATION: The ventilation that occurs naturally in cold housing.

NAVEL CORD: The cord by which the unborn pig was attached to the uterus and nourished. Sometimes during a farrowing the cord may be severed before the pig clears the birth canal; the pig can suffocate, or the cord can twist about the pig's neck to produce the same result.

NEEDLE TEETH: Also called wolf teeth, these are two large teeth on each side of the upper jaw that are present at birth. If not clipped off — not crushed — with side cutters or the like, they can be used to savage the sow's udder or other pigs, and infection will result. Some old-timers leave the small pig in a litter with its wolf teeth to give it a bit of an edge in the fight for a teat. Nursing order will be established in the first 48 hours of life or so, and, once attached to a teat, a pig will return to it.

NIMBY: Not in my backyard.

OFFAL: Waste parts from a butchered animal.

OMNIVORES: Omnivorous creatures eat both animal and vegetable foods.

OPEN: A producer's term for a female of breeding age that isn't bred.

ORAL: Through or via the mouth as a means of treating an animal.

OUT CROSS: The use in the breeding program of an animal that has a completely different genetic background than any animals used previously. In purebred matings, animals with very distinct pedigrees can give their offspring a bit of what might be considered the equivalent of hybrid vigor.

OXYTOCIN: A drug that can stimulate uterine contractions and milk letdown — and one that may well be overused in livestock.

PATHOGEN: An agent that causes disease, such as a bacterium.

PENILE SHEATH: The fleshy sheath that houses and protects the penis. A large sheath can gather urine and semen or injure more easily and become a pocket of infection that reduces breeding performance. This trait seems to breed on and has shown up a great deal more with today's trend toward looser-sided hogs.

PHENOTYPE: How the animal's genetic makeup manifests itself visually. The term covers traits, such as conformation, soundness, and muscling, that are at least somewhat apparent to the naked eye.

PIG: A very young pig.

PLACENTA: The afterbirth that is also expelled at the time of farrowing.

PRESSURE TREATED: Runners, timbers, and structural lumber that have had a wood preservative applied to them under pressure. This forces the preservative deep into the wood to make it last longer.

PRIMAL CUTS: Large cuts of meat, often transported to butcher shops.

PREMIX: A vitamin and mineral mixture to be combined with grain and soybean oil meal to form a complete ration. This is quite often the least-costly means of formulating a ration.

PRESTARTER: A feed very rich in protein and milk sugars to be offered as a first feed to baby pigs.

PROBIOTICS: Beneficial organisms similar to those already present in the gut that can be given to pigs and hogs to further improve gut activity and feed utilization. They may also be used to restore gut activity in sick and recuperating hogs.

PRODUCTION COSTS: All of the costs required to produce a feeder pig or pound of pork. These should include taxes, interest, insurance, and dozens of other costs that can be overlooked in the planning stages of an enterprise.

PUREBRED: This term describes animals that have been bred true for many generations and have traceable pedigrees.

RIDGLING: A male that retains one testicle within its body.

RING: A metal ring that is crimped into the tip of the nose or across its end to keep the hog from rooting. When the pig tries to root, the ring causes it some discomfort.

ROTATION: This term can refer to the use of different breeds in an established order in a crossbreeding program. It can also refer to the movement of hogs to clean pastures or lots, to let previously used land lie fallow to break disease and parasite cycles.

RUPTURE: A protrusion of a part of the intestine through an opening in the wall of muscle surrounding it. It's the same as a hernia.

SCOURS: The swine-producer's term for diarrhea.

SCROTUM: The saclike affair that carries the testicles. Semen is very heat sensitive — so much so that even the boar's body temperature can impair it — and it is the scrotum's purpose to both protect the testes and hold them close to or away from the body to protect the sperm as air temperatures warrant.

SEEDSTOCK: The source of new swine stock.

SEGREGATED EARLY WEANING (SEW): In this practice, pigs are weaned early and moved to a nursery facility far from any other hogs as a method of breaking the disease cycle.

SELECTIVE BREEDING: Putting together matings of animals strong in a particular trait and those weak in that trait to improve that trait in their offspring.

SELF-FEEDING: Allowing the animal to meet its own nutritional needs from a feeder that stores large amounts of feedstuffs and allows the animal unlimited access to them. This is also known as free-choice feeding.

SERVICE: An act of mounting and ejaculating is said to be a service by the boar.

SETTLE: When a sow conceives, she is said to settle.

SET TO THE LEG: This term describes front legs with good position at the corners of the body, cushioning in the feet, and a slope that supports and cushions the hog while it walks, stands to eat, and breeds.

SHED TYPE: A type of housing with a slanting roof extending beyond both the high front wall and the lower rear wall. Doors and openings will be in the front wall, with some means of warm-weather ventilation in the rear.

SHOAT: A term used for a hog from weaning to 120 pounds.

SLAUGHTER CHECK: The examination of carcasses at the time of slaughter for clinical signs of illness. This type of examination is most generally conducted by a veterinarian.

SNARE: A device for restraining hogs that fits around the snout.

SONO-RAY: A device that uses ultrasound to measure lean muscling in a live hog.

SOW: A female that has borne young.

SPECIFIC PATHOGEN FREE (SPF): These are pigs taken from their dams by cesarean section and raised in isolation. The intention is to rear breeding animals free from certain health problems, such as sarcoptic mange. The approach was developed more than 30 years ago in Nebraska. These pigs are used to repopulate or reestablish herds that have had problems with disease. The SPF pigs are similar genetically.

STAG: A hog that has been castrated after reaching maturity.

STARTER: A second-stage feed for the young pig, given from the time it is 14 days old until it weighs 40 to 50 pounds. These feeds have a crude protein content of 16 to 18 percent but are still rich in sugars and milk proteins.

STRESS: A strain or tension on the animal's well-being caused by weather, moving, weaning, changes in ration, castration, or other environmental or physical disruptions.

SUPPLEMENT: Feedstuffs to be added to grains to improve protein content and quality, and to form a complete, balanced ration.

SWINE: A generic term roughly equal to (but a little fancier than) hog.

TERMINAL: A hog used as a cross to maximize growth and muscling, since the pigs produced are intended to be sold for butcher stock.

TESTOSTERONE: A male hormone that is injected into a boar to stimulate sexual activity and libido. Ideally, it needs to be used only once or twice to get a slow-breeding boar started.

THROUGH THE GUT: This is the route preferred by many for treating a sick animal. Oral drugs, feed and water additives, and probiotic products all work via the gut to improve the animal's well-being in the way its systems work naturally.

TOP-DRESS: To place an additive or treatment on top of the hog's regular ration for consumption at the same time.

UNDERLINE: The bottom line of the hog's body on which are presented the penile sheath, udder segments, and teats. It is important that the teats be large and evenly spaced for best presentation and function.

VULVA: The external part of the female genitals. This is a good indication of reproductive tract size and potential productivity.

WALNUT: A small abscessed node.

WET SOWS: Sows that have only just recently weaned pigs and still have distended udders.

WITHDRAWAL: The time frame in which hogs must be withheld from drugs or treatments before being marketed for slaughter for human consumption.

WOLF TEETH: The same as needle teeth; see above.

INDEX

Note: Page references in *italic* refer to illustrations or photographs; those in **boldface** refer to tables.

OTHER STOREY TITLES YOU WILL ENJOY

The Family Cow, by Dirk Van Loon. Practical, fully illustrated chapters with accurate information on buying, behavior, nutrition, breeds, handling, milking, calving, and growing feed crops. 272 Pages. Paperback. ISBN 0-88266-066-7.

Storey's Guide to Raising Chickens, by Gail Damerow. An informative book that enables both beginning and experienced chicken owners to be successful raising chickens. 352 Pages. ISBN 1-58017-325-X.

Storey's Guide to Raising Llamas, by Gale Birutta. A comprehensive handbook covering behavior, training, facilities, showing, health care, breeding, and birthing. 304 Pages. Paperback. ISBN 1-58017-328-4.

Keeping Livestock Healthy: A Veterinary Guide to Horses, Cattle, Pigs, Goats & Sheep, by N. Bruce Haynes, D.V.M. Provides in-depth tips on how to prevent disease through good nutrition, proper housing, and appropriate care. Includes an overview of the dozens of diseases and technologies livestock owners need to know. 352 Pages. Paperback. ISBN 0-88266-884-6.

Small-Scale Pig Raising, by Dirk Van Loon. Information on penning and handling, health and nutrition, commercial feeds, breeding, physiology, and butchering. 272 Pages. Paperback. ISBN 0-88266-136-1.

These books and other Storey books are available at your bookstore,
farm store, garden center, or directly from Storey Books,
Schoolhouse Road, Pownal, Vermont 05261, or by calling 1-800-441-5700.
Or visit our Web site at www.storeybooks.com